HEADLINE NEWS SCIENCE VIEWS II

Edited by
DAVID JARMUL
National Research Council

National Academy of Sciences
National Academy of Engineering
Institute of Medicine
National Research Council

NATIONAL ACADEMY PRESS
Washington, D.C. 1993

NATIONAL ACADEMY PRESS • 2101 Constitution Avenue, N.W. • Washington, D.C. 20418

The National Academy of Sciences is a private, nonprofit, self-perpetuating society of distinguished scholars engaged in scientific and engineering research, dedicated to the furtherance of science and technology and to their use for the general welfare. Upon the authority of the charter granted to it by the Congress in 1863, the Academy has a mandate that requires it to advise the federal government on scientific and technical matters. Dr. Frank Press is president of the National Academy of Sciences.

The National Academy of Engineering was established in 1964, under the charter of the National Academy of Sciences, as a parallel organization of outstanding engineers. It is autonomous in its administration and in the selection of its members, sharing with the National Academy of Sciences the responsibility for advising the federal government. The National Academy of Engineering also sponsors engineering programs aimed at meeting national needs, encourages education and research, and recognizes the superior achievements of engineers. Dr. Robert M. White is president of the National Academy of Engineering.

The Institute of Medicine was established in 1970 by the National Academy of Sciences to secure the services of eminent members of appropriate professions in the examination of policy matters pertaining to the health of the public. The Institute acts under the responsibility given to the National Academy of Sciences by its congressional charter to be an adviser to the federal government and, upon its own initiative, to identify issues of medical care, research, and education. Dr. Kenneth I. Shine is president of the Institute of Medicine.

The National Research Council was organized by the National Academy of Sciences in 1916 to associate the broad community of science and technology with the Academy's purposes of furthering knowledge and advising the federal government. Functioning in accordance with general policies determined by the Academy, the Council has become the principal operating agency of both the National Academy of Sciences and the National Academy of Engineering in providing services to the government, the public, and the scientific and engineering communities. The Council is administered jointly by both Academies and the Institute of Medicine. Dr. Frank Press and Dr. Robert M. White are chairman and vice chairman, respectively, of the National Research Council.

Library of Congress Cataloging-in-Publication Data

Headline news, science views II / edited by David Jarmul; National Academy of Sciences . . . [et al.].
p. cm.
Includes index.
ISBN 0-309-04834-6
1. Science news—United States. 2. Science—Social aspects—United States. 3. Technology —Social aspects—United States. 4, Health—United States. 5. Medical policy—United States. I. Jarmul, David. II. National Academy of Sciences (U.S.) III. Title: Headline news, science views 2.
Q225.H45 1993
303.48'3—dc20 93-16471
CIP

B-099

Printed in the United States of America

Contents

3

THE ENVIRONMENT

5

DIET AND NUTRITION

6

TECHNOLOGY AND TRANSPORTATION

7

THE ECONOMY

8

INTERNATIONAL AFFAIRS

9

LOOKING TO THE FUTURE

10

THE SCIENTIFIC ENTERPRISE

AFTERWORD

INDEX

All of the articles and author affiliations in this book appear as originally published.

Editor's Note

The Information Revolution. The AIDS epidemic. Space travel. In countless ways, changes involving science and technology are reshaping our lives. Computers are transforming our economy. New technology brings us everything from Scud missiles to MTV videos. Medical breakthroughs help us live longer even as global warming and atmospheric ozone depletion threaten our future.

It is nearly impossible to read through a newspaper without finding several stories involving science, technology and health care. But for many Americans, especially those without a technical background, these topics often are confusing, even intimidating. People read conflicting claims about an issue and wonder where the truth lies. They sense their lives being changed by everyone from the farmer in the Amazon to the computer hacker next door. But real understanding remains elusive, hidden in a shroud of jargon and details.

This book will help everyone — expert and non-expert alike — to make sense of some of today's most important issues involving science, technology and health care. The authors include dozens of the world's most prominent experts, writing in a readable and engaging journalistic style. The articles are similar in format to those in the first edition of *Headline News, Science Views*, published in 1991.

As with the first volume, the articles in this edition appeared originally on the editorial and opinion pages of daily newspapers. They were distributed by the National Academy Op-Ed Service. Begun in 1983 under the auspices of the National Academy of Sciences, National Academy of Engineering, Institute of Medicine and National Research Council, the Service provides more than 300 newspapers with timely articles by scientific and technical experts. The papers receive the weekly articles free with exclusive rights within their cities. Among those that have published stories from the service are *The Atlanta Constitution, The Baltimore Sun, The Boston Globe, The Chicago Tribune, The Cleveland Plain Dealer, The Detroit News, The Houston Chronicle, The Miami Herald, Newsday, The Philadelphia Inquirer, The San Francisco Chronicle* and *The St. Louis Post-Dispatch*.

The wonderful cartoons and drawings in this volume were used by editors at subscribing newspapers. The artists and editors granted us permission to reprint the illustrations here.

The Op-Ed Service would not exist without the continued support and encouragement of the newspaper editors who have helped us bring these complex scientific and technical issues into the arena of public debate. We also are indebted to hundreds of study committee members, staff officers and others within the Academy who have shared their expertise and offered advice on story ideas. The entire staff of the Academy news office supports the Service in many ways. In particular, Stephen Push, director of the office, and Patricia Worns, the copy editor, played an invaluable role in producing the articles presented here.

Our greatest thanks is reserved for the authors, who took time out from busy schedules to prepare these articles without pay and under tight deadlines. Making the transition from scientific text to newspaper prose was not always easy, but it was made much smoother by authors whose prominence was matched by their patience, eloquence and genuine desire to reach out beyond the scientific community to the American public.

David Jarmul

1

SCIENCE AND SOCIETY

Science and Pseudo-Science

Carl Sagan

If you go and teach, as I sometimes do, in a kindergarten or first grade classroom, you find a room full of scientists. Their questions are deep and fundamental: Why is the sun yellow? What is a dream? How deep can you dig a hole? They have no idea that there is such a thing as a dumb question.

If you teach a class of high school seniors, you find something completely different: enormous concern about social approval by peers and great wariness about asking "dumb questions."

Most human beings early develop a set of scientific attitudes. It combines curiosity, a sense that the world is knowable, and — slowest to develop — the skepticism of a prudent used car purchaser. Yet something happens between first and twelfth grade — and it isn't just puberty. The natural interests of all these little scientists are replaced by science anxiety.

The inherent curiosity of these young people does not disappear. Yet where science is poorly taught, missing in the national intellectual discourse, not discussed casually over dinner in the average household, and absent from television talk shows, people's natural instinct for it is unsatisfied. They turn somewhere else.

To a significant degree, they turn to superstition and pseudo-science. To take one example, consider the issue of ancient astronauts.

Millions of books have been sold on the thesis that many of the ancient monuments of the human species — e.g., the pyramids of Egypt — were built by extraterrestrials or under extraterrestrial tutelage. The argument, often, is that the author can't figure out how such things were made, so the only alternative is that beings from elsewhere did it.

Now this is a fascinating idea, I agree, but archaeologists have actually studied the subject. They understand how you build pyramids — barges up the Nile, inclined planes, tens of thousands of slaves, a few decades, and you're there. It is a mystery only to those ignorant of archaeology.

Where archaeologists have insufficiently explained their findings, ancient astronauts slip in. I claim that most of the people fascinated by claims of ancient astronauts would be as fascinated by the story of the real monument-builders and how we know about them — if this were accessibly communicated, not just in books but on television.

The same is true of UFOs. I'm personally very interested in the possibility of life elsewhere. I'm not opposed to our being visited; it would save me a whole lot of trouble. We send spacecraft to other planets, and try to find pathetic little martian microbes, of which there seem to be none. Large radio telescopes are being used to see if anyone is sending us a message. I would be delighted if those guys were coming here. But extraordinary claims require extraordinary evidence — and the evidence — real, non-anecdotal, physical evidence — is crummy.

The UFO ecological niche is filled partly because discussions of the nature of scientific evidence are virtually absent from popular culture. Skepticism is present in the used car lot, but absent before the "newspaper" offerings of the supermarket checkout line. It's much more interesting to hear that there *is* extraterrestrial life than to hear, well, we are just at the beginning of our studies and we don't yet know.

People are hungry for science and technology explained in a non-intimidating way with some of the grandeur and wonder left in. I believe the American media and a whole lot of others have grossly underestimated the intelligence and interest of the general public. This is probably true across the board, but it's certainly true in science. We can turn this matter around through the mass media and improvements in the schools.

It is clear there is a problem. We have a society exquisitely dependent on science and technology, in which the average person understands hardly anything about science and technology. This is the clearest imaginable prescription for disaster — especially in a purported democracy. A fascination with pseudo-science is a dangerous foundation on which to base decisions about the environment, health care, defense and the many other urgent problems the nation and the planet face.

September 30, 1990

Carl Sagan *is the David Duncan Professor of Astronomy and director of the Laboratory for Planetary Studies at Cornell University.*

* * *

Confronting Creeping Complexity

Robert W. Lucky

Feeling overloaded?

Many of us are, and not only from eating too much at holiday parties. Fax machines, cellular telephones, electronic mail, voice mail, telephone answering machines, phones in airplanes, pagers, and other devices have us drowning in messages and phone calls. Computers bombard our lives with more information than we can absorb.

Listen to the groan of people programming their VCRs or read best-sellers like *Everything I Needed to Know I Learned in Kindergarten,* and one sees this anxiety of Americans about the stress of modern life. Complexity is a fundamental residue of the Information Age and it is rising steadily — in technology, business, social systems and the daily rituals of life.

It's a trend that deserves more serious attention. The telephone network was easily understandable and manageable only a decade ago. Now it has slipped beyond the comprehension of any single person. The collapse of a significant portion of the AT&T network a year ago underlined the vulnerability mired in this complexity.

Other large interconnected systems are found in transportation, the air traffic control system and the military. Computers that contribute to these systems also provide tools to control them, but one of the most important problems of our time is whether we human beings can manage such extraordinary complexity successfully.

As an engineer who has helped develop the technologies of the Information Age, my own view is that our species is up to the task of managing even a bewildering level of complexity.

That is an optimistic view, and an experience I had recently made me painfully aware of how out of touch it may be with that of other Americans. I appeared as a guest on a television talk show about the future. After speaking glibly about a world made more pleasant by robots, high-definition television and the like, I was roundly criticized by the other guests, who insisted that the world's prospects are bleak.

The environmentalist on the show was strident in his recitation of statistics on pollution. The educator spoke of the decline of literacy. The economist talked about global starvation, and the former police officer sitting beside me on the sofa warned of the inevitability of drugs and crime. When I held to my viewpoint that technology would make the future better, the others looked at me with scorn. What does a technologist know about such things?

That's a reasonable question for Americans to ask of people like me, since we produced this technology and have a dubious record of predicting its impact. Few of the engineers who developed the videocassette recorder imagined that every town today would have a video rental store. The inventors of optical disks concentrated on video applications, never guessing that compact audio discs would displace vinyl records.

So technology produces complexity and is unpredictable, yet engineers like myself remain optimistic about its application. As a consequence, we make progress where none is expected. Unaware that cities are a hopeless cause, we design successful urban transportation systems like BART in San Francisco or the Washington Metro. Oblivious to the hopelessness of the educational crisis, we pursue technological aids to education.

This single-minded pursuit of solutions may be hopelessly naive for the world of the future, and there's no question technology can produce bad outcomes as well as good ones. But I do think most Americans would be better off if they shared our approach of viewing technology as an ally in a world of creeping complexity rather than as the enemy.

Technology and simplicity are not mutually exclusive. In fact, I believe technology increasingly will free us to focus on matters more worthy of our human intellect, producing a world in which art, religion, music and philosophy coexist with amazing technical advances. Technological products are only tools, and they can be used to make life less, as well as more, stressful. The real solution to our frazzled lives lies not with rejecting technology but with harnessing it in new ways to manage information overload, quiet the beepers, and calm our nerves. We need to retain faith — not so much in technology as in our own power as humans to make it work for ourselves.

December 16, 1990

Robert W. Lucky *is executive director of the Research Communications Science Division at AT&T Bell Laboratories.*

* * *

The Reality Beyond Science

Victor F. Weisskopf

From Patriot missiles to new medicines, our lives are affected as never before by developments in science and technology. As one who has dedicated his life to science, I share the widespread concern of my peers about how little many Americans understand about science. Yet I think an opposite problem also exists — people depending too much on a scientific view of the world at the exclusion of other perspectives.

An anecdote about two of my former mentors illustrates the dilemma. The story goes that Werner Heisenberg and Felix Bloch, two of the giants of modern physics, were walking along a beach on a sunny day. Bloch was explaining a new way of looking at the geometrical structure of space. Heisenberg, his mind drifting into complementary avenues of experience, interrupted: "Space is blue and birds are flying in it."

Like Heisenberg, the rest of us also need to guard against seeing the world solely through scientific eyes. When we admire a sunset, for example, we may think of the scientific reasons why the sun is red. Yet we also should be impressed by the beautiful color combinations. The starry sky on a beautiful night may make us think about the fascinating facts of stellar evolution. But it is also symbolic of awe-inspiring infinity and the mysteries of the night. A Beethoven sonata contains not only sonic vibrations but also an emotional message.

Unfortunately, many people, and not only some scientists, resist this complementary view of things. There is a trend today toward clear-cut, universally valid answers. The scientific approach often is considered the only reasonable and serious one. There are good reasons for this belief; no field of human experience seems inaccessible to scientific comprehension. Yet even if scientists eventually unravel the processes underlying our own thoughts and emotions, other methods of discourse still will be needed to grasp complex experiences.

History has shown that whenever one perspective is emphasized, others are unduly neglected. Just as science and technology now are in the forefront, religion dominated during the Middle Ages. The religious view of life was so dominant that nobody in Europe made a scientific record of the appearance on July 4, 1054, of a supernova ten times brighter than the brightest star. In China, by contrast, a detailed record of the phenomenon was made.

The one-sided emphasis on religion during the Middle Ages produced great art and lasting moral principles, but it also led to serious abuses, such as the murderous Crusades and the neglect of corporal suffering. One must ask whether today's emphasis on science and technology has led to dangerous exploitation of our environment, an overemphasis on material values and irrational overproduction of weapons of destruction.

I do not believe a direct cause-and-effect relationship exists. But I do think it important as we approach a new century that we recognize the totality of human experience. For moral and political decisions, scientific insight can point out the consequences of certain actions. The decisions themselves, however, must rest on non-scientific arguments, applying concepts that deal with the human soul and moral values. Applying varied approaches does not mean choosing among them to justify all actions. Nor is it a denial of values. Rather, ethical principles are strongest when derived from many sources and then applied firmly to the situation at hand.

Science and technology comprise some of the most powerful tools for solving problems. But they provide only one path toward explaining the varied and apparently contradictory ways of the mind when we are faced with the realities of nature, our imagination and human relations. Other paths are equally needed, if for no other reason than to help prevent thoughtless and inhuman abuses of science.

Niels Bohr, the father of quantum physics, taught me many years ago that complementary ways of perceiving the world are essential to understanding it. He was right. Now, decades later, we can look forward to a future full of unfathomable scientific and technical advances. Yet we will continue to confront nationalism, religious fundamentalism and other intolerant views. Our only chance of achieving a better world is for all of us — scientists and non-scientists alike — to apply many different approaches rather than relying on science alone.

March 17, 1991

***Victor F. Weisskopf** is Professor Emeritus of Physics at the Massachusetts Institute of Technology. This article is adapted from his autobiography,* The Joy of Insight *(Basic Books).*

* * *

Columbus Day and the Frontier of Exploration

Edward C. Stone

Five hundred years ago tomorrow, three ships commanded by Christopher Columbus made landfall in the Bahamas after a 10-week voyage across the Atlantic. This opened an era of exploration of the Americas by European adventurers and was hailed by generations of students as the discovery of the New World.

Although some historians point to evidence of earlier visits by Europeans, and others are troubled by the cultural and economic consequences that resulted from the meeting of civilizations from two hemispheres, Columbus' feat remains significant. His voyage dramatically extended the frontier of the world as known to his European contemporaries.

Five hundred years after Columbus' voyage, the world that

Drawing by Gus D'Angelo
The Peninsula Times Tribune, Palo Alto, Calif.

once seemed so vast has shrunk to what has been called our global village. The passage between the continents that required weeks of hazardous sea travel is now accomplished comfortably by air in only hours. Satellite television coverage of news in other countries has brought immediacy to once remote events.

The launch of the first satellite, *Sputnik*, by the Soviet Union 35 years ago opened another new frontier. The Space Age expanded rapidly because of the competition during the

Cold War, but it also recaptured the spirit of exploration. Like Columbus' voyage, the journeys into space presented unknown hazards and challenges that we couldn't be sure we could meet. But these journeys also presented opportunities for observing the universe and for exploring and understanding not only Earth but also the other bodies in the Solar System.

The human frontier now encompasses the Moon, even if only briefly so far, and our robotic emissaries have visited all of our planetary neighbors except Pluto. A television series that begins airing on PBS tomorrow night, "Space Age," recounts the remarkable history and examines the future of the space program.

The last 35 years have given us confidence that the frontier of space is accessible. Indeed, space missions often seem deceptively routine, even though they remain among the most challenging endeavors we can undertake. Because weightlessness and other aspects of space cannot be completely simulated in an Earth-based laboratory, space systems can be tested only partially before launch. Once in space, repair is much more difficult or even impossible, yet in many instances the systems must function nearly perfectly for a decade or more.

The bounty of scientific knowledge that our missions have returned could give the impression that there is little interesting left to discover, that the veils obscuring nature's larger secrets have been stripped away, leaving only the details to be observed. I do not believe this to be the case, and I fully expect the century ahead to hold discoveries just as astonishing as those we have already witnessed.

As in the past, some of the most important discoveries will be those that differ the most from our expectations. Columbus anticipated making landfall in the Far East and, in fact, sent a delegation to seek the court of the Mongol emperor when he reached Cuba. He did not foresee coming upon an unimagined continent.

Similarly, in planning the *Voyager* mission to the outer planets, we at the Jet Propulsion Laboratory had no idea that the two spacecraft would glimpse erupting volcanoes on Jupiter's moon Io when they flew past the planet in 1979.

I believe it is invigorating for a society to share a sense that there remains something new to discover and understand, a sense of a future that differs from the present. Space will remain a limitless frontier that can provide the whole world with such a sense, much as Columbus' voyages of exploration provided for the Old World.

Looking ahead to the next 500 years, some of our greatest accomplishments are certain to occur in space. Human exploration will extend to Mars and beyond, while robotic systems will explore regions of space inaccessible to humans, and telescopes on the Moon will observe the distant universe with increasing clarity.

Recent findings about small bodies beyond Pluto, and about fluctuations in the Big Bang and the origin of the Universe, remind us that space provides unlimited opportunities for discovery and surprise. As we commemorate this historic anniversary, the best way to honor Columbus is not with parades but with the renewal of our own human spirit of exploration.

October 11, 1992

***Edward C. Stone**, director of the Jet Propulsion Laboratory in Pasadena, Calif., is vice-chair of a National Academy of Sciences committee that advised the PBS television series "Space Age."*

2

EDUCATION

Children and Calculators

Kenneth M. Hoffman

Forget the upcoming presidential election or Madonna's new movie. The next time you want to pick an argument with someone, tell them our country's schoolchildren should be using pocket calculators more often to learn mathematics.

I've said so to people I've met on airplanes and at parties, and they often look at me like I'm crazy. "If children use calculators in class," they sputter, "how will they learn the multiplication tables?" Or, "Students will know how to push the buttons but won't understand the underlying mathematics." Then they tell me about the time they went to the store when the cash register wasn't working and the teenage cashier didn't know how to make change.

The very idea of using calculators in classrooms hits a vital nerve in many Americans. They view it as cheating and fear that our already dismal level of mathematics performance will worsen.

As one who has spent a lifetime teaching mathematics, I disagree profoundly with these criticisms. There is no evidence that the average young cashier today is any worse at arithmetic than teenagers were 50 years ago, although the growth of the service sector does make their inadequacies more obvious. Teenagers of the past depended on a pad and pencil instead of on a cash register. In practical terms, what's the difference?

The real problem with calculators, I think, is that many Americans view mathematics as something painful that youngsters must study because it's good for them. If Mom and Dad spent countless hours doing long division problems, then, by God, Jason and Kimberly can, too. Such attitudes explain why our students perform so miserably. They have been led to view math moralistically rather than as a liberating tool for understanding the world. Mathematics is seen as a test not only of brains but of character, of whether someone has the grit to calculate problems day after day, year after year. No wonder people hate it.

Calculators can change this equation. Students still must master the basic skills, but now they can escape the drudgery of endless repetition and do new and exciting things. Elementary school students, for instance, can use calculators and other tools to explore subjects currently reserved for higher grades, such as geometry.

Suppose youngsters spent as much time learning about volume and area, by pouring liquids from one container to another, as they now devote to long division. They could discover that a cylinder holds enough liquid to fill three cones with the same base and height as the cylinder. They'd find that three spheres hold just enough liquid to fill two cylinders that have the same radius as the spheres and a height equal to the spheres' diameter. Centuries ago, Archimedes said these relationships are among the most profound truths of nature. Why shouldn't our students have the chance to discover them as well?

Similarly, children should be using blocks and tiles to learn that doubling the sides of a square results in an area four times as great. For older children using calculators, it then is a short step to learn about fractals, chaos and other topics that go far beyond the clerk-training curriculum now in place.

Young children can learn about statistics by measuring the heights of their classmates. A teacher then can guide them to consider ways of determining the center. Is it the average of the heights, the height in the middle, or the height that occurs most frequently? Calculators make it possible to assess these possibilities quickly, keeping students focused on the big picture.

Contrary to many people's assumptions, mathematics is not an unchanging body of facts and procedures. It is the language of science, and it evolves continually. When chalkboards were introduced in schools many years ago, some teachers feared children would lose the ability to write. Modern worries about calculators are likely to prove similarly groundless.

Technology is not a panacea, as many school systems have learned with computer-based learning materials and other reputed innovations. Dedicated teachers and sound pedagogy remain essential. Yet, used appropriately, calculators can make the job easier, and we should not fear them. They give students what their parents lacked: time and freedom to become better problem-solvers and to discover the beauty of mathematics.

June 9, 1991

***Kenneth M. Hoffman**, professor of mathematics at the Massachusetts Institute of Technology, also is associate executive officer for education at the National Research Council.*

* * *

One Year Down, Nine to Go

Timothy H. Goldsmith

Nine years to go. That's how long U.S. students now have to become first in the world in mathematics and science achievement.

At their "education summit" last year, President Bush and state governors set the year 2000 for students to accomplish this goal. Their challenge came as Americans have been debating educational reforms ranging from lengthening the school year to requiring better teacher-competency tests.

Yet American students have little hope of catching — much less surpassing — their counterparts in Japan, Britain and elsewhere until something more fundamental happens. Our science classrooms must stop being so dreadfully boring and irrelevant.

I chaired a committee of the National Research Council that recently examined the most widely taught science subject — biology. We found that biology is taught so poorly in the United States that it frequently snuffs out student interest in all science. Four of every five students take biology in high school, but fewer than a third of them continue with science by studying chemistry. Less than half of chemistry students, in turn, take high school physics. We cannot possibly meet our deadline of the year 2000 with this commitment.

Dismal enrollment figures do not tell the whole story. A recent analysis found that fully half the students who never take a class in biology do *as well or better* on biology tests than 40 percent of their peers who took and passed a biology class. In other words, many students learn almost nothing in these courses — except to dislike science.

Biology classes should be helping students develop an interest in the world around them and an understanding of societal issues such as AIDS and the environment. But from the early grades to high school, biology education is hampered by poorly trained and supported teachers, irrelevant curricula, inappropriate standardized tests, and textbooks that often are inaccurate or misleading. Considerable evidence suggests that the same problems exist in other science classrooms.

Biology and the rest of science need to become an essential part of elementary school, while the curiosity of children is uninhibited. Instead, most students receive little science instruction until junior high school, and then with curricula and textbooks that typically are exercises in memorization rather than an intellectual voyage of exploration.

What's needed is a curriculum that emphasizes science as a process of understanding. It should be an open-ended game of "What if . . . ?" rather than a stupefying exercise in memorizing terminology reinforced by pedantic standardized tests. Students need to confront their beliefs about cause and effect, and spend time experimenting with their own hands and eyes, in simple laboratory settings.

Teachers need better training and should stop being so constrained by bureaucratic directives, non-educational duties and insufficient time to prepare class materials. They also need many more opportunities for meaningful, in-service experiences in which they can interact with research scientists, upgrade their knowledge, and share ideas and experiences.

From a national perspective, the most important need is to keep sight of these day-to-day realities in the classroom as we consider other reforms. Most of the "solutions" that have received attention lately are largely managerial or administrative. Examples include lengthening the school day or the school year, requiring more multiple-choice tests to establish the competence of teachers or the progress of pupils, opting for alternatives to the traditional certification or licensing of teachers, and adjusting the relative authority of teachers, principals, superintendents and boards of education. A voucher system, allowing parents to choose their child's school, is the latest such enthusiasm.

Some of these approaches contain useful ideas. But few specify or assure the kinds of fundamental change that are most necessary to improve learning, those that must take place in the classroom. Significant improvement among U.S. students in science cannot occur without better training for teachers, more relevant curricula and textbooks, and tests that focus on concepts rather than terminology. These fundamental reforms are the first step to young Americans' excelling in science — and in the complex world of the future.

Only 15 percent of high school seniors now opt to take science, so students are voting with their feet about what's wrong. With just nine years to go, we should listen to them.

December 30, 1990

Timothy H. Goldsmith *is professor of biology at Yale University.*

* * *

Minority Students and Mathematics

Asa G. Hilliard III

Millions of minority students graduate from high school unable to do simple equations, evaluate statistics or perform other mathematical tasks essential to success in today's technological society.

This isn't news. But what amazes me is that even though this problem is widely lamented, we hear little about teachers across the country who have shown absolutely that it doesn't need to be this way.

Jaime Escalante, the man who inspired the movie "Stand and Deliver," is the only "success story" known to the public or, for that matter, to many mathematics teachers. He helped minority students at Garfield High School in Los Angeles to excel in calculus.

Just across town from Garfield, a group of regular third graders at the Marcus Garvey School in Watts triumphed in a math competition over a group of sixth graders from a magnet school for gifted students. In the Bedford-Stuyvesant section of Brooklyn, a regular fifth grade class received special instruction for one year and then passed New York's ninth grade math examination in overwhelming numbers.

I had a similar experience in Denver, helping a class of poor white and Hispanic students to excel in algebra two years before they were even supposed to study the subject. My purpose here is not to suggest that Hollywood update "Stand and Deliver" with an epic about any of these experiences. Damage already has been done by well-meaning tales that lead people to conclude that minority students can learn mathematics only if they have a teacher with superhuman dedication.

Since few teachers, or others, are so dedicated, such a conclusion is tantamount to saying that many students will continue to fail in math and have little chance to become computer programmers, pilots, engineers or accountants. That conclusion is simply wrong. Even teachers with low expectations of their students can be trained to improve student performance dramatically.

The record shows how this can be accomplished. One lesson we can learn from Escalante and others is to bypass the basics and focus on intellectual challenges. Uri Treisman, who helped African-American students in Berkeley to achieve dramatic increases in mathematics scores over a short period, went straight to calculus. Bill Johntz, who has worked with students in Texas and California, taught fifth graders some concepts usually introduced during the freshman year of college. The students did not feel overwhelmed, but challenged.

This strategy is exactly the opposite of what many public schools do. When kids get behind, the schools pull them aside, slow them down, and feed them "dumbed down" courses, bit by bit, at a slower pace. This strategy is guaranteed to destroy children's enthusiasm.

Successful teachers achieve results quickly. Renee Wilkerson Anderson in Portland, Ore., is among those who have helped minority students improve performance rapidly. She and the other teachers all had a strong personal background in mathematics. They encouraged students to ask questions, introduced them to real mathematicians, and strove to develop a feeling of competency among young people who previously had not experienced much success.

One thing these teachers did not emphasize is grades. Nor did any of them divide children into ability groups. Although the vast majority of American schools use grouping, a growing body of empirical evidence indicates this approach doesn't work. Also contrary to prevailing wisdom, these teachers had few special materials, textbooks or equipment. Bill Johntz and Uri Treisman used little more than chalk and a blackboard. Elementary school students at the Garvey School studied from ordinary college algebra textbooks.

These cases also show that the best way to train new teachers is through an apprenticeship. A big problem in education today is that few prospective math teachers ever see success as a reality rather than an abstraction. They need to watch master teachers in action.

The contrast between these teachers and others is striking, and it suggests that dramatic changes are needed in mathematics education not only for minority students, but for all young Americans. Successful pedagogy is not special. It is reproducible and can work for minorities and everyone else. We should stop regarding excellence in mathematics as anything other than the norm.

November 11, 1990

Asa G. Hilliard III *is the Fuller E. Callaway Professor of Urban Education at Georgia State University.*

* * *

A Failing Grade for School Tests

Jeremy Kilpatrick

If you've gone to school, you've probably endured at least one teacher who "taught to the test," spending day after day preparing you for a state biology exam, the SAT or some other standardized test. A class like that can be a monotonous, miserable experience for teacher and student alike, crushing the spontaneity out of learning.

Why then are so many politicians and educators calling for "new standards" as a way to improve our troubled schools? President Bush, for one, said at the National Academy of Sciences that "we can't expect kids to meet the test of worldwide competition unless we first establish world-class standards that define the knowledge and skills we expect students to learn and master."

Won't new standards put even more pressure on teachers to rigidly follow a prepared curriculum that turns off their students? As one who has been active in mathematics education reform, that is a concern I hear frequently from teachers, parents, students and others.

The concern is understandable — but misplaced. The real problem is not the existence of standards and tests; it is that we are using the wrong standards and the wrong tests for the wrong reasons. Tests should measure what our society truly values. All too often, they now measure what is easy to test.

In most mathematics classes, for example, tests largely measure computation and routine procedural skills rather than a student's ability to apply mathematics in the real world. Multiple choice and short-answer tests, with their emphasis on quick responses, are used excessively. These tests do not allow students to show how they can integrate and apply their mathematical knowledge.

Consider these two test questions:

Question One. What percentage of 500 is 30?

A. 6% C. 60%
B. 16.7% D. 166.7%
E. None of the above

Question Two. In 1980 the education budget of a certain community was $30 million out of a total budget of $500 million. In 1981 the education budget of the same community was $35 million out of a total budget of $605 million. The inflation rate for that one-year period was 10%. Do the following tasks:

A. Use the facts to argue that the education budget *increased* from 1980 to 1981.

B. Use the facts to argue that the education budget *declined* from 1980 to 1981.

The first question might have been adequate at a time when society was content with young people mastering routine mathematical skills. But as we have learned so painfully in recent years, our economic and social well-being now requires a population with greater thinking, reasoning and learning skills. We no longer can treat mathematics as a set of recipes to be memorized and, all too often, forgotten.

Our national experience with the Advanced Placement exams taken by many college-bound high school students shows it is feasible to create more flexible assessments that include portfolios of student work, essays and other free-response tasks. Other approaches might involve the use of student debates, computer demonstrations or simulation models. The same argument extends beyond mathematics. A good way to assess students in science, for example, is to ask them to develop a hypothesis and conduct a simple experiment.

Classroom teachers are in the best position to make these evaluations, and we should place more trust in them. Teachers need help improving their own assessment methods, not pressure to conform to commercially developed tests of questionable relevance.

Many standardized tests are misused. The SAT and ACT college entrance exams, for instance, were developed to help make good matches between individual students and colleges. But now they are cited widely — and inappropriately — as measures of the performance of school districts, states or the entire country. School systems should be evaluated directly, not with dubious extrapolations from the tests of college applicants.

Better assessment methods obviously cannot transform American education by themselves; establishing tough new standards without also providing the necessary instruction and resources is just setting up students for failure. We also should be wary of adopting a single national curriculum or test, which would raise many other problems.

Still, the basic fact is that we cannot make our schools more accountable unless we state clearly what we expect students to achieve — and then measure whether they have succeeded. Doing so is not "teaching to the test." It's common sense.

March 8, 1992

***Jeremy Kilpatrick**, professor of mathematics education at the University of Georgia, chaired a group studying mathematics assessment for the Mathematical Sciences Education Board of the National Research Council.*

* * *

Getting Scientists Involved in Science Education

Ramon E. Lopez

Millions of young Americans barely know the difference between a protein and a proton. In a world that depends on science and technology, they're in big trouble. The irony is that the United States possesses the world's most productive scientific community — many thousands of people blazing a path in immunology, astrophysics and other fields.

An obvious question is why more of these experts don't help students in local elementary, junior high and high schools to overcome their ignorance of science and become the world's best in the subject by the end of the decade, as President Bush has proposed.

Scientists teaching kids about "real science" might work wonders. Students could hear for themselves how exciting it is to unravel the mysteries of diseases or distant galaxies.

Unfortunately, although scientists complain about science education regularly, they tend to be like most people in not getting involved in something that doesn't affect them directly. Once at a scientific meeting I asked everyone to sign a volunteer list for local schools. One of my colleagues rolled his eyes and said something to the effect of, "Oh, no, Lopez is at it again." Needless to say, he was not interested in helping.

Another reason scientists aren't doing more is that they may come to a school expecting to "fix" a situation they do not really understand. Well-meaning scientists sometimes believe all educational problems would be solved if only the teachers would listen to them. They fail to recognize that knowing something about chemistry or biology does not make them experts in teaching young people.

Furthermore, some scientists have a poor opinion of teachers and difficulty treating them as equals. The teachers I have known are dedicated, hard-working and intelligent. Given innovative materials and the necessary training and resources, they do an excellent job. What they need is not condescension but support.

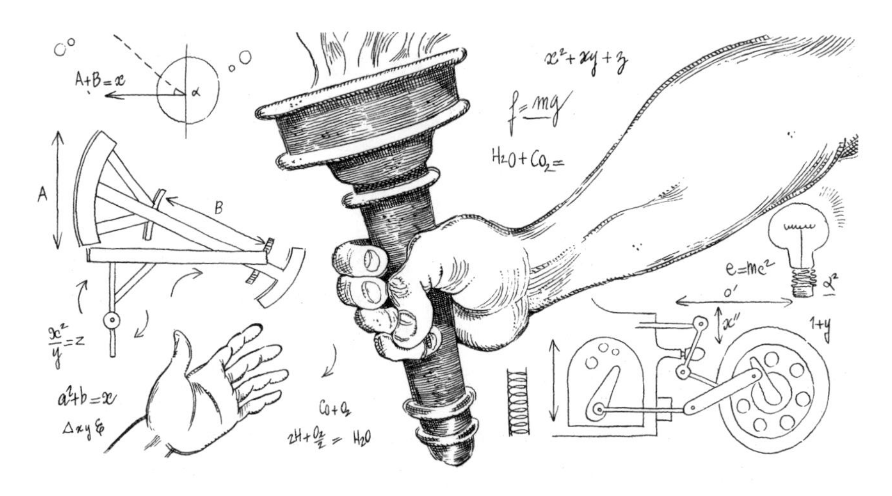

Drawing by Ben Dib
The San Bernadino Sun, Calif.

Perhaps the biggest underlying problem is that many scientists continue to see science education as a filter for identifying a handful of interested people like themselves rather than as a pump that injects everyone with excitement about science. Such an open approach is especially important with girls and minority students, who now are badly outnumbered in the sciences and must fill the ranks in the future.

Rather than just listening to lectures and memorizing facts, young people should be encouraged to make hypotheses, observe, measure, and draw conclusions on their own — in other words, to learn science by doing science instead of just reading about it. Even if they don't become scientists, they must know how to evaluate facts and make judgments, skills essential in a democratic society.

Although several obstacles impede scientists from becoming more active, the fact is that a growing number are working hard to help improve local schools. Several scientific organizations have organized efforts to help them.

The National Academy of Sciences and the Smithsonian Institution, for example, have established a National Science Resources Center, which has begun holding workshops that train scientists to assist at schools. The center also brings together scientists and teachers in developing innovative teaching materials, such as kits for experimenting with electric circuits or for cultivating fast-growing plants. Programs elsewhere are helping scientists share their skills in ways students can understand and emulate.

Even scientists who receive this training face a problem. They generally work long hours and survive on grants and contracts of short duration. They must write proposals and publish results. There are few nights and weekends left over for personal projects such as working in local schools. I have heard many colleagues say, "You know, I'd really like to help but just can't find the time."

The heart of the problem, in other words, is not a lack of good intentions but of resources and structure. Scientists could provide schools with expertise and role models to inspire students to raise their sights to the heavens. But they cannot do it alone. They need guidance on how to be truly helpful. They need employers to support their efforts. And they need schools that really want to change. Scientists and parents alike should be demanding reforms like these to produce excellence in science education. Otherwise, our children may grow up to be scientific dimwits in an increasingly scientific world.

July 26, 1992

***Ramon Lopez**, a space physicist in the astronomy department at the University of Maryland, College Park, has worked closely with local public schools.*

★ ★ ★

Barbie, Math and Science

Mildred S. Dresselhaus

A new Barbie doll that says "math class is tough" has been greeted with hoots of derision for teaching girls to fear math and science. As one of the relatively few women of my generation who grew up to succeed in the scientific world, I am glad to see this reaction. Our daughters shouldn't have to overcome the same hurdles we did.

My career opportunities were created largely by the advent of *Sputnik* in 1957, the year before I received my doctorate. Constant encouragement from my mentor, Rosalyn Yalow, played an important role in my early career.

I am glad to see the manufacturers of Barbie dolls being taken to task for perpetuating the tired old myth that girls cannot excel in math and science. Barbara McClintock, who just passed away, was among the greatest geneticists of this century. Gertrude Elion — like Yalow and McClintock, a Nobel laureate — helped revolutionize the way we treat diseases. Susan Solomon is an authority on atmospheric ozone depletion. Mae Jemison was a physician and engineer aboard the last space shuttle mission. The list goes on and on.

Still, for all of the progress they have made, American women hold only 16 percent of our country's science and engineering positions. That is not enough. The "Barbie controversy" reminds us how much more women could be contributing to curing diseases, cleaning up the environment, modernizing factories and uncovering the secrets of nature. Discouraging girls from pursuing math and science not only cheats them as individuals but also squanders one of our country's most precious resources — the brain power of half its citizens.

A recent survey by the American Association of University Women found that most girls between the ages of 8 and 18 have a negative view of math and science and of their ability to perform as well as boys in these subjects. Their self-perception is affected significantly by the behavior of their parents and teachers.

Simply providing girls with more enlightened Barbie dolls will not overcome this situation. To encourage more women to pursue careers in science and engineering, we need intervention programs that help women overcome the many barriers that still exist.

At the pre-college level, much can be learned from Harvey Keynes at the University of Minnesota. He has implemented changes in recruitment procedures, course curricula and teaching styles that have resulted in more girls pursuing science studies as they prepare for careers, citizenship and parenthood. Science is an important part of our cultural heritage.

At the undergraduate level, universities need to make women feel welcome in physics, chemistry, engineering and other classrooms where they may be outnumbered by males. Women should be encouraged to participate actively and not be treated as intruders. I have found through my long career on the MIT faculty that many women benefit from peer group and networking programs that provide women with information, guidance and reassurance that they can succeed.

"Big Sister" and other mentoring programs can provide women with role models that help them visualize their own scientific success. At the University of Washington, for example, undergraduate and graduate female students are matched with scientists and engineers on the faculty and in the surrounding community. Other universities have programs that help women science and engineering students improve their communication skills.

In the business world, companies need to recruit more women for technical positions. Those already hired should band together to provide mutual support and to overcome the organizational barriers that keep "glass ceilings" in place.

To succeed, these and other efforts need active support from the top, whether on campus or in the corporate suite. Institutions also must become more adaptable in helping scientists and engineers balance their jobs and family responsibilities. Too many laboratories now cling to a macho work ethic that leaves no room for children. This tradition also hurts fathers, but is especially difficult for women.

Interventions like these are essential if American women are to help our country deal with global warming, the AIDS epidemic, economic competitiveness, and a host of other problems involving science and technology. Ken cannot handle all of these problems alone; we need Barbie, too. Setting her straight about math is only a start.

October 18, 1992

***Mildred S. Dresselhaus** is Institute Professor of Electrical Engineering and Physics at the Massachusetts Institute of Technology and treasurer of the National Academy of Sciences. She chairs the National Research Council Committee on Women in Science and Engineering.*

* * *

The Contrast Between Computers and Classrooms

Kenneth G. Wilson

Over the past thirty years, I have lived in two worlds, the world of computers and the world of teaching and learning. The contrast between them is stark. In the 1960s, I could only dream about computers fast enough to help me unlock secrets of atoms and nuclei. Within a decade, supercomputers became central to my research. Today, a personal computer, or PC, can outdo the supercomputers I used in 1970.

My ability to teach students, however, advanced hardly at all. I became frustrated as millions of children failed to learn the basics of mathematics and science. Even if I abandoned my research and taught full time, I could personally help only a tiny fraction of them.

Why has progress come so rapidly to computers but so slowly to education? One important reason is that computer designers have put in place a system for monitoring how their machines are faring, seeing what works, and making improvements quickly. PCs were not designed by a few employees at a local computer store on their lunch break.

Teachers, by contrast, have been expected to plan for reforms in their classrooms with little outside help. They are stretched thin just trying to help a subset of their students.

We need to learn from this contrast. Although many worthwhile efforts are under way to improve our schools, they will be transitory and insufficient unless the *process* of change is incorporated into education. Well-intentioned bursts of reform cannot produce the steady progress achieved with computers.

A number of pioneering organizations already have helped 50 or more schools with reform planning or professional development. Among them are several "whole school restructuring" programs, such as Robert Slavin's "Success for All" organization at Johns Hopkins University. Others are mentioned in the book *Smart Schools, Smart Kids*, by Edward Fiske.

Erie, Pa., has an especially interesting project — a 10-year effort to apply the principles of W. Edwards Deming to local workplaces and schools. Deming's ideas of "quality management" played a large role in modernizing industry in Japan.

These and other organizations are analogous to the young computer designers at Apple Computer who began designing the Apple II even while the first Apple computer was enjoying success. The computer experts had the capital to undertake a new redesign cycle. Educational reform groups, which are non-profit, do not.

As a result, progress in education remains momentary. We take one step forward and then slip back to the usual lethargy. Good ideas in one school are unknown elsewhere.

I have two recommendations at the federal level to improve matters. The first is to set aside a modest percentage of federal educational budgets to enable successful education reform programs, such as Slavin's, to broaden their scope to include ongoing redesign.

Second, the National Science Foundation should organize

a major new initiative on this issue of evaluation. It should support needed research and provide large numbers of promising recruits from disadvantaged populations with multi-year traineeships to evaluate reforms in their local schools. Since these populations have the greatest unmet needs, they should be involved directly in attempts at improvement.

Education reform cannot be a one-shot effort. There must be continuing experimentation with ideas such as students teaching each other in small groups and as "teacher's helpers." Materials need to be redesigned repeatedly. So does software for inexpensive, hand-held PCs. Teachers need sustained professional development programs in their schools to keep up to date on subject matter and teaching methods.

My main recommendation for readers is to join a community reform movement, such as the one in Erie. If none exists, learn about Erie's program and talk it up in your community. Be heard. Write your newspaper or member of Congress.

Obviously, improving schools differs in many ways from redesigning computers. Children are not microchips. Yet the opportunity of linking the two worlds is too stunning to ignore. The concept of a redesign cycle has helped to revolutionize communications, transportation, agriculture and other fields. The opportunity now exists to demonstrate that the same process can achieve accelerated progress in education. No nation has attempted such an adventure. If the United States wants education reforms that are sustainable, it must lead the way.

December 6, 1992

Kenneth G. Wilson*, a Nobel laureate in physics, is a professor of physics at Ohio State University. He helps direct "Project Discovery," a program to improve mathematics and science education in middle schools throughout Ohio.*

* * *

Fooling Ourselves About Improving Ourselves

Robert A. Bjork and Daniel Druckman

Many people who would never dream of buying a car without checking a consumer magazine become amazingly incautious when it comes to methods that purport to improve human performance.

Americans spend more than $50 million annually, for example, on "subliminal learning" tapes that claim to help listeners with problems ranging from losing weight to building self-esteem to becoming a better bowler. There are meditation classes to reduce stress and self-assessment tests to guide career decisions.

Having just directed a National Research Council study that evaluated a variety of techniques designed to enhance performance, we advise readers to be selective in their enthusiasm.

There is neither a theoretical basis nor empirical evidence, for instance, that subliminal self-help audio tapes can alter complex human behaviors. Many people believe they have been helped by such tapes, but the available research suggests that any changes for the better are due to processes such as "expectancy effects," when a person is so ready to change that it matters little what's on the tape.

There also is no convincing evidence that meditation has any special properties as a technique to reduce stress and control tension. Rest and relaxation training appear to be as effective. People who meditate regularly may pursue a more peaceful lifestyle, but one must distinguish between the practice of meditation and these lifestyle changes to determine why stress or tension was reduced.

The popularity of self-assessment tests as a tool in career counseling also seems unjustified. One such test, the Myers-Briggs Type Indicator, classifies people into certain "personality types" and is administered to nearly two million people in the United States each year. Yet our committee could find no convincing evidence of a relationship between Myers-Briggs types and performance in particular occupations.

So caution is needed. At the same time, however, there is good reason for optimism about some techniques. We concluded, for instance, that people can be taught psychological techniques to help manage pain. Proven stress management techniques such as relaxation training, providing information about what to expect, and enhancing a person's sense of control all can help people cope with pain.

There also is evidence that some mental rehearsal and preparation techniques are effective in helping performance. Mental practice can be useful when learning a motor skill, and mental rehearsal of a learned skill can facilitate getting ready to perform. Simple rituals such as bouncing a tennis ball a certain number of times before serving may slow a person's heartbeat and produce other physiological changes associated with better performance.

Conventional training techniques also need closer scrutiny. Many training programs offered by companies, for example, compress instruction into a short period, fix the conditions of practice, and provide continuous feedback to the students. These approaches may facilitate performance during training, but they are not effective in terms of long-term retention or applying skills to new situations.

What works better is to space practice sessions over time, vary the conditions of training, and provide feedback only intermittently. Such measures introduce difficulties during training, but they result in more durable and flexible skills after training is completed. Training that fosters understanding and involves students as active participants in the learning process also improves long-term performance.

Why would corporations that spend billions of dollars annually to teach workers to operate machines, use computers, and carry out other tasks, use inefficient methods? And why do experienced instructors use these methods? The answer, in part, is that instructors usually see students' performance only *during* training and even may be evaluated themselves by that performance. On-the-job performance needs to become the main criterion by which training programs are evaluated.

Our human drive to improve ourselves is laudable, but less admirable is our tendency to believe there are easy solutions to difficult problems. It's time to stop kidding ourselves. Dramatic claims and testimonials, even when accompanied by good intentions, are not enough. We have to be guided by hard evidence. Yes, it is possible to improve how we learn and perform. Those improvements can even be dramatic, but they are rarely effortless.

November 3, 1991

Robert A. Bjork *is professor of psychology at the University of California, Los Angeles.* ***Daniel Druckman*** *is a study director at the National Research Council.*

* * *

The Overselling of the University

Lester C. Krogh

The ivy-covered walls are growing a different kind of green these days. Increasingly, American universities are selling their wares.

Corporations are making arrangements with universities for patentable ideas and trained workers. Federal and state governments are turning to them for extension services and to develop new ideas for industry. Foreign governments and companies are reaching out to our campuses for early access to research findings and for training their students.

Many universities, in turn, are aggressively marketing an ever-broader range of services. The Georgia Institute of Technology advises local businesses through its Industrial Extension Service. Worcester Polytechnic Institute's Manufacturing Engineering Applications Center develops products for subscribers. Stanford University and the University of California at Berkeley maintain active industrial affiliate programs. Similar examples abound.

Many of these arrangements provide valuable educational opportunities for college students. Yet if universities become too eager in their pursuit of new revenues, they could lose sight of their main mission — the training of the next generation. That would be disastrous not only for universities and students, but for all of us.

As Princeton University president Harold Shapiro has pointed out, universities only recently have been expected to make a dollars-and-cents contribution to economic growth. Over the past eight centuries, their main product has been their graduates, who go on to influence the economy through their daily working lives, reshaping society without fanfare. The day-to-day job of education is less glamorous than campus research that wins Nobel Prizes, but it represents technology transfer at its most profound and lasting.

There are many reasons why university administrators have begun looking beyond this traditional mission and marketing new services. They are struggling with post-baby boom enrollment declines, rapidly rising administrative and facilities costs, and shrinking pools of government support. Most universities also sincerely want to help government and business make better use of good ideas developed on campus.

Nonetheless, at least some universities are now in danger of becoming victims of their own sales pitches. They endlessly cite a few notable successes — Silicon Valley in California, Boston's Route 128 and Research Triangle Park in North Carolina — as evidence of the economic leverage of their own proposals. These marketers risk becoming mercenaries if they advertise too direct a relationship between higher education and higher profits.

For professors, the pursuit of new sources of research funding may be the inevitable outgrowth of a "justify your existence" mentality. The vicious academic cycle of "publish or perish" puts pressure on them to constantly write new research proposals — or write up their résumés. This mindset has helped make teaching careers so unattractive that U.S.-

born professors are now a rarity in some disciplines, particularly in science and engineering. Teaching has taken a back seat to research because it simply doesn't pay for universities or professors. This shift away from teaching threatens to prevent our daughters, sons, employees and other students from getting full value from their education — and from our education dollars.

Make no mistake; university research is essential for generating new ideas, discoveries and technologies. But it is not a sure-fire ticket to prosperity. The openness of our university system is essential to intellectual vitality. Yet it also ensures that research findings, in many cases funded by our government, can be picked up easily by foreign companies. There is no guarantee that the benefits of university research will remain in the United States. We can be much more certain that our students — the real product of our colleges and universities — will invest their careers in our country's economy.

So, as they seek to meet rising expectations with declining resources, universities should be temperate in their promises of economic return. And companies and governments must avoid raiding universities for their intellectual breakup value. We can enhance our national industrial competitiveness only with careful planning, patient effort and hard work. We should expect no magic potions from campus labs.

The real return on our personal and collective investments in universities is the career-long contributions of our graduates. We cannot put too high a value on their training, and we must not forget that the university's focus should be on people, not profit.

February 24, 1991

Lester C. Krogh *is retired senior vice president for research and development at 3M Corp. He chaired a National Academy of Engineering symposium on the relationship between universities and economic development.*

3

THE ENVIRONMENT

The Threat of Climate Change

Daniel J. Evans

While Operation Desert Storm grabbed everyone's attention, another potential crisis was forgotten. The combined pressures of population growth and pollution are threatening our planet's resources and could change our climate. We must now pay more attention to this other threat, which has the potential to cause environmental damage far more widespread than occurred in the Gulf.

In 1989, Congress asked the National Academy of Sciences, the National Academy of Engineering, and the Institute of Medicine to evaluate both the scientific facts and policy implications of global climatic change. I chaired a group of 40 distinguished scientists, economists and policymakers that has just completed this study.

We found that amounts of greenhouse gases which trap the sun's radiation and help warm our Earth, are increasing rapidly. These gases come from burning fossil fuels, harvesting our forests, and the escape of modern chemicals into our atmosphere. Human activity soon will push concentrations of these gases to levels unprecedented in human history.

Virtually all scientists agree on these facts. Wide differences begin to appear when scientists project what will happen to the Earth's climate as a result of these added greenhouse gases. Computer models concur that a doubling of current pollutants will cause global warming. Yet the projected amount of warming varies from 2 degrees to 9 degrees Fahrenheit. Some scientists disagree, saying there is little evidence that any significant global warming will occur.

Policy-makers have been left wondering how to proceed. After all, a temperature increase of 9 degrees is much more serious than a 2 degree rise — or no rise. Yet, despite this uncertainty, the rationale for taking action is compelling. The situation is much like facing a potential earthquake or fire. One does not know when a disaster will occur, or how damaging it will be. A prudent person buys insurance.

We discovered an array of such insurance measures that could substantially reduce pollutants in the atmosphere. Many of these measures cost little; some actually save money. Collectively they represent an insurance policy for the planet that we all should be eager to buy.

First, we should ensure that consumers pay the full environmental costs of various energy sources. The marketplace can lead us to the least polluting energy sources — if we let it. Research needs to be speeded up on alternative energy sources and also on a new generation of safe, economical nuclear power plants. If the intensity of global warming increases, nuclear energy may be an important alternative to coal- and oil-fired power units.

The main component of a planetary insurance policy should be improved energy efficiency. Everything from better home insulation to more efficient light bulbs and refrigerators will reduce energy use. Best of all, the price of energy saved will pay for most of the improvements.

Many of these measures have multiple benefits. They not only diminish the threat of global warming but also reduce acid rain, smog, and pollution of our air and water. Perhaps the best example of a double benefit is stopping the use of chlorofluorocarbons, which damage the Earth's protective ozone layer and contribute substantially to global warming.

Our committee identified a series of low-cost proposals that could reduce U.S. emissions of global warming gases collectively by an impressive 40 percent of 1990 levels. Adopting these ideas at even a quarter of proposed levels would cut our current pollution load by 10 percent.

Greenhouse warming is potentially a global problem. Even if the United States adapts to climate changes, other nations could face catastrophe. It is vital to share new technologies and improve energy efficiency abroad. Otherwise, our own efforts may be overwhelmed by growing industrialization in the developing world.

There is a lot of talk about a new world order. Here is a pretty good place to start. Helping others to grow in an environmentally sound way contributes to world stability while avoiding climatic disaster.

No one can predict yet how significant global warming will be, but we need not wait for a clearer signal. Adopting the insurance policies outlined here will produce much benefit for little cost. The real bonus is that they also will make us more efficient and competitive, and create a healthier environment for our children.

April 14, 1991

***Daniel J. Evans**, a registered engineer and chairman of a Seattle consulting firm, is the former governor and U.S. senator from the state of Washington.*

* * *

Designing a Cure for Greenhouse Warming

Thomas H. Lee

Recently some scientists have proposed ingenious — some would say reckless — techniques to bring global warming under control and prevent the Earth from overheating.

According to one suggestion, more than 50,000 large mirrors would be launched into orbit to reflect sunlight before it heats our atmosphere. A similar proposal would send billions of shiny balloons aloft.

Another idea is to deliberately release huge amounts of dust into the atmosphere to block sunlight. Then there is the possibility of releasing sulfur dioxide above the oceans to stimulate cloud formation — still another way of blocking sunlight. According to one estimate, increasing the coverage of marine stratocumulus clouds by just 4 percent would offset the effect of a worldwide doubling of carbon dioxide (CO_2) emissions.

Yet another approach is to fertilize the oceans with iron to stimulate the growth of aquatic plants and hasten the absorption of CO_2 by the oceans.

Some of these proposals have an aura of science fiction, and one need not be an engineering whiz to fear they might cause inadvertent environmental problems worse than those they solve. Technological fixes also might lull society into complacency about global warming.

Even though there is evidence of global warming in the last century, whether it is due to greenhouse effects still is uncertain. But because the potential repercussions may be severe, we need to examine all mitigation options, including "geoengineering ideas" like these. I chaired a panel of the National Academy of Sciences, National Academy of Engineering and Institute of Medicine that has just completed an exhaustive study of ways of mitigating global warming. Our conclusion was that these options warrant further study, although none are close to being ready for actual implementation. But we also found that there are other options that are less costly, less exotic, and perhaps even boring by comparison — but that can work.

For example, if homeowners replaced just three incandescent light bulbs in each household with high-efficiency fluorescent tubes and purchased more efficient refrigerators and water heaters, they could reduce residential electricity demand substantially. If everyone also used a more efficient car, the United States might reduce its emissions of greenhouse gases by as much as 15 percent. Even though the financial paybacks of these improvements are very attractive, implementation might require creative policies.

If companies switched to alternative chlorofluorocarbons (CFCs) to replace those depleting the ozone layer, the amount of potential global warming would ease by a similar magnitude.

Though no single silver bullet exists, it is possible to reduce U.S. emissions of greenhouse gases substantially in a number of ways. Our panel divided these options into two categories. The first group includes those available at little or no cost. Several, such as improving the energy efficiency of homes and commercial buildings, actually provide cost savings. The list also includes making power plants more efficient and collecting gas from landfills.

The second category includes options that may cost more or have implementation obstacles. For example, increasing the use of natural gas for electricity generation can offer both savings and a reduction in CO_2 emissions. But concern about the availability of natural gas is an obstacle. Transportation options such as improving mass transit, parking management or vehicle efficiency beyond certain levels require changes in lifestyle, which may be difficult.

The cost of these options varies, from less than one dollar for every ton of CO_2 emission reduced to more than $500 per ton. Measuring costs solely in terms of avoided emissions, of course, leaves out other critical factors. For example, eliminating CFCs not only eases global warming but also protects the ozone layer.

So the equation is complicated. But after toting up all of the options, we concluded that the United States can reduce its greenhouse gas emissions by between 10 percent and 40 percent of current levels. This is a very impressive amount, although emissions in other parts of the world also must be controlled.

In other words, rather than just worrying about global warming, we have the potential to mitigate it at relatively low cost. And we may not need exotic technological fixes. Mirrors in the sky may be unnecessary if we have the vision to act here on Earth.

July 21, 1991

***Thomas H. Lee** is president of the Center for Quality Management and professor emeritus at the Massachusetts Institute of Technology.*

★ ★ ★

Aquatic Ecosystems on the Critical List

John J. Berger

From San Francisco Bay to Lake Apopka in Florida, many of the nation's aquatic ecosystems are on the critical list.

These rivers, lakes, streams, estuaries and wetlands have been dredged, channelized, diked, dammed, massively diverted, silted in and contaminated. Left in shocking condition, they are losing native plant and animal species as well as their capacity to perform such life-sustaining ecological functions as absorbing wastes, purifying water and producing oxygen. Currently, the nation is destroying 290,000 acres of its wetlands every year, not to mention other resource losses.

Confronted by increasing population pressures and relentlessly growing demands on water resources, our aquatic ecosystems urgently need restoration.

An expert committee of the National Research Council concluded recently that it is indeed possible to repair damaged aquatic ecosystems to a close approximation of the condition they were in before they were disturbed. The committee recommended a coordinated national program of aquatic ecosystem restoration to rehabilitate 10 million acres of wetlands, 2 million acres of lakes, and 400,000 miles of rivers and streams.

The newly emerging science of restoration ecology makes such an ambitious goal increasingly realistic by providing tools to improve the condition of drained bottomland hardwood wetlands, channelized rivers, fishless streams and dying lakes. Specific resources that might benefit from restoration include the Chesapeake Bay, the Great Lakes and Midwest "prairie potholes."

Restoring these areas would not only help the environment but would provide much-needed jobs for those involved in the effort, as well as providing potable water, fishing, swimming, and other recreation and tourism. Other benefits include improved flood control, water quality and ground-

Drawing by Brian Duffy
The Des Moines Register, Iowa

water supplies, and increased numbers of fish and waterfowl. Restoration efforts also can provide a sense of empowerment for community groups that participate.

A significant message of the Research Council report is that when people manage natural resources and plan their restoration, they need to consider the long-term, large-scale interactions among rivers, lakes, streams, wetlands and groundwater, and the impacts of land-use practices on water systems.

Current restoration efforts, although worthwhile, have tended to be narrow in scope and uncoordinated on a regional basis. Many government agencies have limited jurisdiction and are only able to manage water quality or water quantity. Restoring an entire ecosystem, however, requires a much broader geographic perspective and a long-term approach that takes full account of the relationship between interconnected water resources and between those resources and their surrounding lands.

With myriad social needs tragically unmet, can we afford to pay for a national environmental restoration program?

The question really should be, Can we afford not to? Environmental damage that is not repaired promptly often becomes tremendously expensive or impossible to repair later.

It is unconscionable to pass on these costs to future generations or to foreclose future options for them. Once environmental degradation gets to a certain point, nature may be unable to repair the damage on a meaningful time scale. Endangered species cannot wait for help in the long-term future. If we do not act to restore their habitats now, they will be gone from the Earth forever.

Public and private spending on the environment for all purposes is a mere 1.6 percent of the gross national product, and a tiny fraction of that amount is being spent today on ecological restoration of aquatic ecosystems. Additional money for restoration must be found. Such a program could be part of the nation's economic recovery effort.

With national attention now focused on ways of revving up the economy, we must be wary of quick economic fixes that ignore vital environmental restoration needs. Narrowing the legal definition of "wetland" without careful scientific planning, for example, could cause unintended harm. We should be moving instead to restore degraded resources so we achieve a net gain rather than just the "no net loss" that President Bush has proposed for wetlands.

In addition to all of its domestic dividends, a bold national environmental restoration program could confirm the United States as a world environmental leader. It also could give us a head start on what ultimately will be a vast enterprise in the next century: global environmental restoration. Desirable as it is, environmental protection is not enough. We also must fix the environmental harm we have done.

February 9, 1992

John J. Berger*, professor of environmental policy at the School of Public Affairs, University of Maryland, College Park, served as special consultant to the National Research Council's Committee on Restoration of Aquatic Ecosystems.*

★ ★ ★

Science and the National Parks

Paul G. Risser

They are some of our country's brightest jewels, but Yellowstone, the Grand Canyon, Acadia and other national parks have a serious problem. The parks face growing pressures, from haze that obscures scenic views to a spiraling number of visitors. Yet those managing the parks are failing to take full advantage of one of their best tools for overcoming these problems — namely, science.

When oil spilled onto Alaska's Kenai Fjords and Katmai National Parks after the 1989 *Exxon Valdez* accident, a lack of data about pre-spill conditions made it impossible for park officials to fully assess the damage. A similar lack of understanding led to concession stands and other facilities being built in inappropriate places in Yellowstone and Yosemite, and to park boundaries failing to encompass complete ecosystems in Everglades National Park.

There are some research success stories. In Sequoia and Kings Canyon National Parks, managers believed for years that they should always try to prevent fires. Research showed that occasional fires are actually required for new trees to germinate; the policy has now been changed. In Great Smoky Mountains National Park, studies showing a decline in the black bear population led officials to uncover illegal hunting.

Scientific research is essential if the park system is to continue serving more than 250 million visitors annually while preserving precious resources for future generations.

Nearly a dozen independent reviews over the past 30 years have found the Park Service's research effort to be poorly organized and inadequate. The Service devotes much less of its budget and personnel to scientific research than do other agencies that administer federal land. The national parks could be producing some of our best research on the environment, geology, archaeology and a host of other fields. Instead, their scientific contributions are sporadic.

The Park Service has failed to respond adequately to these

repeated criticisms. Its scientific program remains plagued by inadequate resources, a vague mandate and a lack of independence even as the need for research has intensified. Air pollution is now obscuring scenery at the Grand Canyon, Shenandoah and other parks. Exotic plants and animals are invading park boundaries. Park managers must cope with critical changes in ground and surface water, increasing stream sedimentation and threats to wildlife populations. The crush of visitors at some of the parks is straining resources to the breaking point.

The park system currently conducts research both on its own and in cooperation with universities. It manages this research effort out of its Washington office, its 10 regional offices and some individual parks. Instead of having a single science program with a coordinated focus, it has 10 separate programs of uneven quality. The Service has identified $250 million to $300 million in unmet research needs, yet its research spending in fiscal year 1992 totaled only $29 million.

Still, the main problem is not money, although more money is needed. Rather it is with the very culture of the Park Service, which treats science as a secondary concern. Glacier National Park, the Grand Tetons and other treasures do not exist solely to impress tourists. They also provide an unparalleled source of untouched natural settings to study evolutionary adaptation, ecosystem dynamics and other natural processes. The parks can teach us about natural and human history, and deepen our understanding of regional and global environmental changes.

What can be done to set things right? I chaired a National Research Council committee that recently studied the situation. We urged Congress to take the lead by issuing an explicit mandate clarifying the Park Service's research mission. The Service, meanwhile, should act swiftly to enhance the program's credibility and quality. It needs to reach out to the broader scientific community by establishing an independent science advisory board and a competitive grants program. It also should recruit a chief scientist of high scientific stature.

The leadership of the Park Service has said it generally

agrees with our recommendations, but it must move decisively to show that a new era truly is dawning.

Incremental change is not sufficient. What's needed is nothing less than a metamorphosis in the Park Service, one that integrates science with day-to-day park management. Our national parks are too valuable an inheritance, and the problems they face too complex, to continue managing them with sketchy information and good intentions.

September 13, 1992

Paul G. Risser *is vice president and provost at the University of New Mexico, Albuquerque.*

* * *

Assessing the Threat of Toxic Waste Sites

Anthony B. Miller

If you live in the United States, there's roughly a one-in-six chance that your home is located within four miles of a chemical dump or other potentially hazardous waste site. Given our unpleasant memories of Love Canal and other incidents, it's reasonable to ask which of these more than 31,000 sites truly pose a threat.

Unfortunately, more than a decade after Congress established the Superfund program, we still cannot answer that question. A committee that I chaired for the National Research Council reported recently that the federal government has no comprehensive inventory of waste sites, no program for discovering new sites, insufficient data for determining safe exposure levels, and an inadequate system for identifying sites that require immediate action to protect public health.

The Environmental Protection Agency (EPA) has conducted preliminary investigations of 27,000 of the reported sites. About 9,000 of these have been studied more extensively, and 1,200 have been placed on the National Priorities List for eventual cleanup. Yet the methods used to assess the public health danger at these sites are questionable, and it is far from clear how much the assessments have benefited nearby residents.

Opinion polls show the public believes that hazardous wastes constitute a serious threat, but many scientists and administrators in the field disagree. Our committee, which included experts in toxicology, exposure assessment and other fields, found the available evidence too skimpy to confirm or refute either view.

More than 5 billion metric tons of hazardous waste are produced each year in the United States. There's no question that substances toxic to humans and several animal species abound in hazardous waste sites. It's a big step, however, to say that most, or many, or even a substantial fraction of the sites pose a threat to nearby residents. Residential proximity does not necessarily mean that exposures and health risks are occurring, although the potential for exposure obviously is increased.

Epidemiologic studies of hazardous waste sites have complex technical limitations. However, increased rates of such conditions as birth defects, spontaneous abortions, cardiac anomalies, fatigue, and neurologic impairment have been tied to exposures among some nearby residents.

It is less clear whether exposure to the wastes can be blamed for medical problems where there is a long delay between exposure and disease. However, some studies have detected excesses of cancer in residents exposed to compounds found at some hazardous waste sites.

As for which sites are a problem or how close people have to live to be affected, much remains uncertain. Without clear answers, the only prudent course is to err on the side of public safety, just as we do in designing bridges or buildings. In evaluating the potential danger of a dump, officials should apply a large margin of safety.

Everyone would benefit, however, by reducing the uncertainty about hazardous wastes generally and about specific sites. Of the $4.2 billion spent annually on hazardous waste sites in the United States, less than 1 percent has gone to study health risks.

The scientific basis for evaluating Superfund sites must be improved. Expanded studies are needed — and soon. As toxic wastes disperse, more people will be exposed and it will become increasingly difficult to design studies that compare the health of exposed and unexposed populations. One kind of research that is especially important is identifying biologic markers that indicate whether someone has been exposed to toxic chemicals.

More broadly, the federal government should establish an aggressive program to discover hazardous waste sites. It needs to revamp its methods for evaluating known sites for population exposures, health effects and the need for cleanup measures. Washington also should expand technical assistance to state hazardous waste programs and increase support for university research in "environmental epidemiology."

After spending billions of dollars during the past decade to study and manage hazardous waste sites, the American people are entitled to firmer information. The only way to end the uncertainty over whether, or how much, sites endanger the public is to perform the necessary studies. We should strive to clear up this scientific mystery even as we clear up the wastes themselves. With more than 40 million people living near the sites, the public needs answers.

January 12, 1992

Anthony B. Miller *is a professor in the Department of Preventive Medicine and Biostatistics at University of Toronto.*

* * *

Protecting Our Nervous System from Toxic Chemicals

Philip J. Landrigan

As you work in your office or walk around your house today you may be exposed to chemical substances that, under certain circumstances, can injure the human nervous system.

Millions of workers are exposed every day to organic solvents. More than 4 million American children have excessive exposure to lead. Other Americans are exposed to pesticides, ethanol, illicit drugs, tranquilizers and other substances with a potential for harm.

Of the more than 70,000 chemical substances used commercially in the United States, fewer than 10 percent have been tested for neurotoxicity. Only a handful have been evaluated thoroughly. Which of these substances truly pose a danger to our nervous systems? We don't know. By using a better scientific system, we could find out more.

Although most substances in our environment are benign, some have been shown to cause neurological illnesses in both adults and children. Examples include:

- Acute lead poisoning in children who ate chips of lead-based paint.
- Severe disfigurement and mental retardation in residents of Minamata, Japan, who were exposed to mercury released into a bay by a plastics manufacturer.
- Tremors, motor disturbances, anxiety and diminished coordination among industrial workers in Hopewell, Virginia, who helped manufacture the pesticide kepone.
- Acute neurologic poisoning by the pesticide aldicarb among California residents who ate contaminated watermelons.

Chronic exposure to some environmental substances also has been shown to cause neurological disease. A syndrome resembling Parkinson's disease develops in people exposed to excessive levels of the metal manganese, as well as in young adults who used the synthetic heroin substitute MPTP. Chronic exposure to lead can cause irreversible declines in children's intelligence and learning, as well as behavior problems. Prolonged exposure to some solvents can produce dementia.

Still unresolved is the question of whether exposure to neurotoxins can result in Alzheimer's disease, parkinsonism or other chronic degenerative disorders of the nervous system. Available evidence suggests that toxic environmental substances may contribute to the incidence of these diseases. The problem is complicated by long latency periods.

A committee of the National Research Council, which I chaired, reported recently that most of the chemical substances to which Americans are exposed in their homes, jobs and hobbies never have been evaluated for possible toxic effects on the nervous system. Accordingly, we called on the federal government to adopt a more effective approach for assessing neurotoxicity to protect the public.

The Environmental Protection Agency (EPA) has the primary federal responsibility for regulating the entry of new chemical substances into the marketplace. Until now, EPA scientists have evaluated neurotoxicity mainly by examining the chemical structure of new molecules. For the most part, they have not done direct testing of toxicity nor have they required chemical manufacturers to do so.

This approach is too roundabout and may fail to identify neurotoxins. EPA should be testing chemicals directly to see whether they are toxic to the nervous system. Proven methods exist to do this.

The agency should adopt a three-tiered testing approach to identify neurotoxic hazards, assess the doses that cause harmful effects, and determine how toxicity occurs. It also should monitor a broader range of chemicals and carry out long-term monitoring on high-risk populations, such as certain workers. Physicians across the country need data to help identify neurologic illnesses linked to toxic pollutants.

Chronic neurological disease has the potential to sap the vitality of American society. Some historians speculate that chronic lead poisoning 2,000 years ago was one of many factors leading to the demise of the Roman empire. Today we live in a society that has synthesized thousands of new chemical substances. Most have enhanced our lives but a few may cause ill effects such as blindness, coma and dementia.

The danger goes beyond possible neurological disorders. Other Research Council committees have reported that environmental substances may harm the immune system, the lungs and the reproductive organs. There is no escaping the world around us but we can be much more vigilant about the substances that can harm us. First, however, we must start doing a better job of identifying which ones they are.

March 22, 1992

Philip J. Landrigan *is Ethel H. Wise Professor and chair of the department of community medicine at Mount Sinai School of Medicine in New York.*

★ ★ ★

Indoor Radon: Hype Versus Help

Anthony V. Nero Jr.

A public service advertisement on television shows people turning into skeletons after being exposed to radon gas in their homes. Print ads proclaim that living in a house with radon is like "exposing your family to hundreds of chest X-rays yearly."

The ads were produced by the Environmental Protection Agency (EPA) and the Ad Council with the best of intentions. High levels of radon pose serious risks of lung cancer. Yet the ads, which aired until recently, alarmed millions of people unnecessarily while inadequately helping those who truly are at risk. The same is true of the federal program behind them.

Radon is produced by the radioactive decay of radium, which exists in small amounts in all soil and rock. Radon decays to isotopes that can deposit in the lungs, causing cancer. Tens of thousands of families are exposed to greater radiation doses than allowed for workers in uranium mines. The increased risk of death is estimated to exceed one in a hundred, or one in ten for smokers.

Most people, however, face a much smaller risk. What the EPA should be doing is vigorously helping those in danger while not scaring everyone else. Instead, it has promoted a mythical picture of indoor radon that substantially exaggerates both the prevalence of homes having high concentrations and the size of the associated risks.

EPA recommends that every home be monitored and remedial action be taken in homes with radon levels above 4 picocuries per liter of air. About 6 percent or 7 percent of U.S. houses exceed this level. The agency has recommended the use of short-term tests that greatly inflate these percentages — to as high as 30 percent of homes — and then publicized the misleading results.

The exaggeration occurs because indoor radon concentrations fluctuate day to day and week to week. Monitoring kits placed for only a few days fail to show the true average. Also, until recently homeowners were advised to place the kits in the basement, where levels tend to be substantially higher than in upstairs living areas.

EPA said the higher levels cause a lung cancer risk equivalent to smoking a half-pack of cigarettes per day, and tried to goad parents into action by saying children suffer even greater risks. Substantial criticism by the scientific community caused it to back off both claims.

In trying to help the public, the agency's staff may feel that exaggerating the situation errs in the right direction. But the approach is backfiring. Americans just don't buy the emergency tone anymore. Even people who really are at risk dismiss the danger.

The United States needs a more sensible and effective control strategy. The essential components are clear: Inform the public reliably. Focus on finding and fixing homes where radon levels are unequivocally high. Examine seriously what, if anything, should be done for the majority of homes with low or moderate levels.

An essential requirement is to adopt a monitoring protocol in which radon detectors are left in a home's main living area for a full year, or long enough to provide an accurate estimate. EPA should intensify its efforts to identify which specific areas of the country have the highest radon levels. Then it should encourage monitoring of every home in these areas, followed by remedial measures where necessary. The usual remedy, at a cost of $1,000 or so per home, is to install a venting system that draws radon-bearing air out from beneath the house before it enters. A moderate but focused program, devoting perhaps $500 million to the 100,000 "hottest" homes, would be much more effective than the unfocused multi-billion dollar programs now being promoted by EPA and Congress.

Real estate transactions in high-radon areas may need to include effective radon testing or remediation, much as many states now handle termite problems. The costs could be distributed widely through new insurance mechanisms. Special building codes also may be required in these areas.

More broadly, EPA must overcome its reluctance to include the scientific community in decisions involving radon. Scientists now must publicly confront the agency for lapses that would never have occurred had they been involved. It's time to stop crying wolf about radon and start getting serious.

November 1, 1992

Anthony V. Nero Jr. *is a senior scientist in the indoor environment program at the University of California's Lawrence Berkeley Laboratory. This article is adapted from a longer version in the Fall 1992 edition of* Issues in Science and Technology.

★ ★ ★

Deciding Who Gets Western Water

A. Dan Tarlock

Prolonged drought is a fact of life in the West, and every day of sunshine illuminates the need to allocate our shrinking water supplies more equitably and efficiently.

The federal government once could build enough new reservoirs to meet the demands of growing cities and the environment while still assuring supplies for agriculture. But the current system is straining to meet the water needs of homeowners in California, fishermen in the Northwest, farmers in Colorado and millions of others.

Irrigated agriculture, long the biggest water user in the West, clearly must relinquish some of the water obtained historically through the doctrine of "prior appropriation." But how?

The best way is through voluntary water transfers, such as an alfalfa farmer agreeing to sell water rights to an expanding suburb. Transferring water from willing sellers to willing buyers has the greatest potential to reallocate supplies fairly and efficiently.

However, there's a problem. A water transfer may make both the seller and buyer happy but harm others. The farmer's neighbors, for instance, might have to pay a larger share of maintaining the irrigation system. The town could lose part of its tax base. The local farm machinery store would have one less customer. Fish habitats and other aquatic ecosystems downstream might disappear. These and other "third parties" may have their lives turned upside down even though they were not involved directly in the transfer.

If water transfers are to succeed in easing growing shortages, Western states must find a better way to protect the interests of third parties such as rural communities and American Indians, as well as of "unrepresented" public interests like the environment. I chaired a National Research Council committee that recently studied the situation. We saw an urgent need for new procedures and laws to promote equity and efficiency. A broader group of participants needs to be included when water transfers are negotiated.

Both urban water suppliers and large segments of the environmental community have embraced the concept of water marketing. Along with many farmers, they see markets as more effective than government subsidies or regulation at moving scarce resources from lower- to higher-value uses. Markets respect property and allow current owners to set the timetable and receive fair compensation.

The concept already has been proven effective in many instances. Most transfers to date have been from farms to cities but they also can go to support ecological systems or recreational uses. Transfers from an irrigation district east of Reno, Nevada, for example, are being used to restore a wildlife refuge.

Water markets differ from conventional markets in important ways. For one thing, water has been publicly subsidized for decades. Large amounts of it currently are held by public and private entities. And we have a long tradition of water supporting a variety of public needs. Selling water rights is not like selling a car or home; the transaction can impose significant costs on others. In some Hispanic villages in northern New Mexico, for example, the entire social system is tied to communally organized irrigation canals.

The ultimate goal of water policies should not be simply to promote transfers but to accomplish better water management generally. This requires that all of the relevant third parties be brought into the deliberations. Although this broad participation may complicate the process during the short term — and increase costs — it helps avoid lawsuits and other problems down the road.

Token gestures to include community groups and others in discussions over water transfers are not enough. Our committee called for third parties to be given legally cognizable interests. Some states have begun to develop new processes to evaluate transfers and accommodate the diverse economic and cultural values associated with water use. But more action is needed — not only by states but on the federal and tribal levels as well. Special attention must be paid to protecting both the regions where water originates and non-consumptive "in-stream" uses such as those involving wetlands or recreation.

Arguments over water rights can be fierce. Voluntary water transfers transacted through the marketplace hold out the promise of bringing adversaries together with deals that are both fair and sensible. For transfers to work effectively, however, everyone with an interest must be seated at the bargaining table.

April 12, 1992

A. Dan Tarlock *is professor of law at Chicago Kent College of Law in Chicago.*

4

HEALTH CARE

A Health Agenda for Children

Frederick C. Robbins

With the presidential campaign moving into high gear, candidates are offering proposals designed to win support from middle class voters, older voters, unemployed voters, Southern voters, New England voters, ethnic voters and many others.

What may be needed most of all, however, is an agenda to help those who cannot vote: our children. Their declining status in society has been oft noted and widely decried, yet they claim a diminishing share of our country's resources. Federal spending on the elderly increased 52 percent during the past decade while spending on children declined 4 percent. One in four children under age 6 lives in poverty.

As one who helped develop childhood vaccines, what especially disturbs me is that so many American children now suffer needlessly from preventable health problems. How can a society that claims to care deeply about its children excuse the recent resurgence of measles or unnecessarily high rates of infant mortality? Our platitudes about loving children are especially hollow when solutions of proven value go unused. We know the following measures all can improve children's health substantially, at reasonable cost:

- ***Provide every expectant mother with prenatal care.*** The United States ranks 24th among the world's nations for infant mortality; one baby dies in every 100 live births. Studies by the Institute of Medicine of the National Academy of

Sciences and other organizations have shown that incidents of low birthweight and neonatal mortality can be reduced with comprehensive prenatal care. Providing basic care to expectant mothers is less glamorous than rushing in later with advanced technology to rescue premature infants, but it saves money, lives and heartache. This is especially true in cases where a pregnancy is complicated by the mother's diabetes, poor diet, alcoholism, drug abuse, age or other risk factors. Every expectant mother should be provided with prenatal care.

• ***Vaccinate all children.*** Vaccination is one of the most cost-effective public health measures. Immunization for diphtheria, whooping cough, tetanus, polio, hemophilus B infection (the cause of meningitis and severe respiratory infection) and hepatitis B should be administered early in life. Although some vaccines have been linked to occasional adverse side effects, their benefits outweigh the risks. In some populations in the United States as many as half of all children have not been fully vaccinated. This rate improves by school age, since all states require children to be immunized before enrolling, but waiting until age 5 is too late. Children need to be vaccinated earlier, and universally.

• ***Enroll all eligible children in Head Start programs.*** President Bush recently announced plans to expand Head Start. That is welcome news, but many children still will be excluded from this worthy program, which helps get disadvantaged children ready for school. Fewer than half of the children who could profit from these services now receive them. The long-term benefits more than justify making them available to all who need them.

Other measures, such as providing health care and counseling in schools, also have been shown to be effective. The Special Supplemental Food Program for Women, Infants and Children (WIC) has succeeded in improving the nutritional status of children and pregnant women substantially. Yet the WIC program now reaches only a fraction of those who need it. Another critical need is to provide adequate health insurance to children, who now account for a disproportionate share of our nation's medically uninsured. In 1989 children comprised 29 percent of the population but accounted for 36 percent of those without health insurance.

"Children are our future" may be a cliche but also is a truism. If we do not act to improve the health of America's children, that future is threatened. As the campaign proceeds and the merits of various health care proposals are debated, we are sure to hear the pleas not only of the candidates but of a great assortment of special interest groups. Somewhere in that din, I hope more voters will be asking, "What about our children?"

March 1, 1992

***Frederick C. Robbins**, winner of the Nobel Prize for research that led to the development of the polio vaccine, is University Professor Emeritus at Case Western Reserve University School of Medicine in Cleveland.*

★ ★ ★

Childhood Vaccines: The Parent's Responsibility

Harvey V. Fineberg

Parents of young children know the dilemma. Their local school probably requires the children to receive recommended vaccinations, but reports have linked some of the vaccinations to adverse reactions.

During the coming year, millions of families will wonder what to do. Having chaired the scientific committee that produced one of the most widely publicized studies of the DPT and rubella vaccines, I would urge virtually all parents to have their children receive both vaccines, as well as other recommended vaccines — against measles and polio, for example.

It is true that a causal relation exists between the DPT and rubella vaccines and certain health problems. Our committee of the Institute of Medicine of the National Academy of Sciences reported that about one in 50,000 children who receive the DPT vaccine suffers from anaphylaxis, a potentially life-threatening allergic reaction. A much higher percentage — as many as six per 100 children — cry inconsolably for several hours shortly after being vaccinated. Weaker evidence suggests a causal relation between the DPT vaccine and two conditions — shock and acute encephalopathy, a brain disorder.

In other words, the DPT vaccine does have the potential to harm some children. But the risks are much smaller than those of the diseases it prevents. The DPT injection gets its name from three such diseases: diphtheria, pertussis and tetanus. All three are potentially deadly and not to be taken lightly. The threat of these diseases far exceeds the potential dangers of the vaccines.

The vaccine for pertussis, or whooping cough, has been the most controversial. Pertussis is a very serious respiratory infection in which the patient typically suffers from a frequent, intensive cough. It can lead to major health complications and death. Although the incidence of pertussis has declined dramatically in the United States since vaccination became widespread, it remains a major cause of child mortality in the developing world.

The number of cases also skyrocketed in Britain and Japan after parents there stopped vaccinating their children because of worries about adverse reactions. In Britain more than 100,000 cases and 36 deaths were reported after the vaccination rate dropped from 80 percent to less than 30 percent.

The continuing problem with measles in our own country illustrates the need to remain vigilant about providing childhood vaccinations. More than half the children in some U.S. communities today have not been vaccinated for measles. In 1989 about 18,000 cases of measles were reported in the United States. In 1990 the total climbed to about 28,000 reported cases and nearly 100 deaths. A decline in vaccination rates must not be allowed to occur with the DPT and rubella vaccines.

Like the DPT vaccine, the shot for rubella, or German measles, has drawbacks. It appears to cause acute and sometimes long-term arthritis in a minority of patients, especially if those being vaccinated are young adults. Yet rubella, too, poses serious threats. Maternal exposure to it during pregnancy can result in numerous congenital health problems for the infant. Protecting young children against this threat is well worth the risk, particularly since side effects of the vaccine are so rare among younger patients.

These vaccines are very good but imperfect. Research into the development of even safer vaccines should continue, but parents can use existing vaccines with confidence that they are doing the right thing for their children.

Parents do need to keep their eyes open to potential problems. Pediatricians should be informed if a child is not feeling well on the day of a planned vaccination; the shots may be postponed. For children who receive the vaccines, some crying is normal. But any signs of swelling or difficulty breathing should prompt a fast call to a physician. Another danger sign is lethargy or difficulty in awakening.

Five years ago the federal government established a program to provide compensation for those children who do suffer severe reactions. Since then, there has been a growing backlog of cases. These families have suffered real tragedies, and the very least they deserve is a faster resolution of their cases.

Their misfortune, however, does not change the fact that the vaccines are saving many more lives than they harm. As one who helped carry out a comprehensive evaluation of their risks, I would not hesitate to urge my own family and friends to get their children vaccinated.

December 29, 1991

Harvey V. Fineberg*, dean of the Harvard School of Public Health, chaired a committee of the Institute of Medicine of the National Academy of Sciences that reviewed the adverse consequences of the DPT and rubella vaccines.*

* * *

The Neglected Mental Health Problems of Adolescents

John J. Conger

From the new television show "Beverly Hills 90210" to movies like "Bill and Ted's Bogus Journey," Americans enjoy laughing about the problems of adolescents. But growing up is not funny for millions of American teenagers who suffer from depression, schizophrenia, drug abuse, anorexia and other serious mental health problems.

The suicide rate among older adolescents has tripled since 1950. The incidence of major depressive disorders among adolescents has risen. In a recent national survey of eighth- and tenth-grade students, 15 percent of the boys and a staggering 34 percent of the girls reported that they often felt sad and hopeless.

This is not to say that emotional turmoil is the norm for American adolescents. Although this period of life has long been recognized for its vulnerability, the great majority of adolescents manage to make the transition to adulthood without major emotional upheaval. But epidemiological studies do indicate that about one in nine suffers from clinically recognizable disorders during any one year.

Despite the magnitude of the problem, adolescents are one of the most poorly served population groups in terms of their mental health needs. Deficiencies extend from research and prevention to social, vocational and clinical services.

Relatively little attention has been given to the differences in mental health problems of older boys and girls. For example, adolescent girls are at much greater risk for eating disorders like anorexia and bulimia. Male adolescents outnumber females in completed suicides, but attempted suicides are more common among females. Although depressive disorders are more frequent among boys than girls before puberty, the reverse is true after puberty. Smoking has decreased dramatically among boys but far less among girls.

What accounts for these differences? Some theorists attribute much of the disparity to biologically based differ-

Drawing by Eleanor Mill
Mill News Art Syndicate

ences in such characteristics as aggressiveness, interpersonal styles, sexual maturation and physical strength. However, biology clearly is not the whole story; differences in how boys and girls are socialized also matter.

For instance, the transition from childhood to adolescence generally is more complex for girls than for boys. Adolescent girls still tend to receive mixed messages about sexuality, vocational aspirations and appropriate behavior. More of what is expected of boys in terms of physical and sexual development or academic, social and vocational expectations represents a continuation of existing trends. For girls, changing expectations often are more abrupt.

The influence on mental health of these factors, both biological and social, still are poorly understood by researchers. Our ignorance about them is tying our hands in efforts to prevent or more effectively treat drug abuse, eating disorders, crime and delinquency, mental and emotional disturbances, and other adolescent problems. It also leads to a great deal of heartache for families. Successful efforts at prevention during these years could yield lifelong benefits.

Until recently, research on adolescent development lagged far behind studies of both younger children and adults. Fortunately, the quantity and quality of developmental research on adolescence has increased significantly in the past decade.

No amount of progress in research, however, will help adolescents avoid mental and addictive disorders unless there is a similar increase in preventive programs and treatment services that put research findings into action. Despite repeated protestations of concern for our children, we are failing miserably as a country to begin to meet their needs. The situation is growing worse, not better. Stresses on adolescents are multiplying even as federal, state and other support for basic mental health and other vital services declines.

Ironically, some adolescents receive expensive treatment, such as hospitalization, while many others remain underserved or not served at all. Because adolescent psychiatric centers are among the most profitable sectors of a financially pressed hospital industry, many centers are marketing their services through television commercials and other ads. Some of these centers offer excellent treatment. But too many adolescents are being hospitalized inappropriately. Meanwhile, increasing numbers of young people who truly need high-quality hospitalization or residential care, but lack the financial resources, are denied treatment.

Our society must go beyond political rhetoric and make a far more serious commitment than it has shown to date to meeting the needs of its children and adolescents. Otherwise, the benefits of even the most imaginative research, prevention and treatment programs will be limited, at best.

August 18, 1991

John J. Conger *is professor emeritus of clinical psychology and psychiatry at University of Colorado Health Sciences Center.*

* * *

People's Health, Public Health

Steven A. Schroeder

Health care is emerging as a major issue, but a big problem with the debate so far has been that people are talking too much about medicine. We should be talking more about public health, which is quite a different question and arguably a more important one.

Medical care is targeted at individual patients; public health deals with the population as a whole. If we finally are getting serious about restructuring our health care system, we should be asking how to help the most people.

The best way to reduce premature mortality from heart disease, cancer, suicide, stroke, injuries, AIDS and many other leading causes of death is to combine traditional medical interventions for individual patients with strategies aimed at society generally.

Although the United States spends more on health care than does any other country, our health indices — such as infant mortality and life expectancy — consistently lag behind those of our economic peers. Put simply, our system does not provide adequate value for the money.

In many countries whose statistics we envy, public health measures are seen as important national health strategies. Tough drunk-driving and gun control laws or contraception and sex education to prevent unwanted pregnancies all are used effectively. In the United States we lack the will to intervene as effectively with preventive measures aimed at populations as we do with measures focused on treatment for individuals.

An important exception is our response to cigarette smoking. Indeed, smoking in North America has declined more than in Western Europe. But this progress has come about in large part from spontaneous citizen action supported by the remarkable leadership of our nation's Surgeons General, from Luther Terry to Julius Richmond to Everett Koop, three men who had the will to promote public health strategies.

The schism between medicine and public health must be healed, but doing so will be especially difficult in the United States. In our country, public health is associated with government and citizens do not now seem to trust that government, or even to respect it.

In recent times, three successful politicians from different parties and very different viewpoints — George Wallace, Jimmy Carter and Ronald Reagan — ran for the nation's highest office on the platform that government is the enemy. Many members of Congress and officials from state and local governments are doing the same. In no other nation is the public health enterprise so identified with unpopular government for which the public is reluctant to pay. This reaction is so strong and reflexive that even a public program that would have provided potentially great benefits — the Medicare catastrophic coverage legislation — was defeated, at least in part, by the very people it would have helped.

A report of the Institute of Medicine of the National Academy of Sciences noted three years ago that millions of Americans would benefit if the nation began taking public health more seriously. That is even more true today; government and the academic community should take the lead in making it happen.

The federal government, for instance, only recently has begun to resolve such contradictions as one agency calling for more generalist and fewer specialist physicians, while another agency pays doctors with incentives that favor specialism over generalism.

At the nation's 126 medical schools, departments of preventive, social, community, or environmental medicine generally are poorly endowed and accorded low status. Nor is there much contact with public health faculty and researchers. Medical schools should expand their curricula and begin teaching not only how to treat low-birthweight babies, but also how to bring about the prenatal care and community services that lead to healthier births. In addition to treating trauma victims, perhaps young surgeons should consider ways of preventing homicides and motor vehicle accidents.

Many individual practitioners still may choose to spend most of their time on either clinical medicine or public health, but the gulf between the two approaches should not be so wide. And if we truly want to improve our ailing health care system, society as a whole needs to bridge this gulf as well. The cures we seek are not to be found only in doctors' offices.

December 1, 1991

Steven A. Schroeder is *president of the Robert Wood Johnson Foundation.*

* * *

The States and Health Care Innovation

Molly Joel Coye

Momentum is growing to overhaul the country's costly and inadequate health care system, but what should take its place? Much attention has been given to systems in other countries, notably Canada, but we should not overlook some intriguing and successful alternatives closer to home.

Many states have experimented with approaches that might be adapted nationally:

• Last year, Hawaii became the first political jurisdiction in the country to offer universal access to health care. It uses a combination of employer-mandated and state-subsidized insurance.

• A year earlier, Massachusetts began implementing legislation requiring employers to "play or pay" — to provide health insurance or pay into a state fund for the uninsured. The plan now appears to be stalled.

• The state of Washington subsidizes enrollment in private health maintenance plans for more than 18,500 of the state's uninsured poor.

- For the past decade, New Jersey has reimbursed hospitals for the full cost of care provided to uninsured patients.

In all, at least 28 states have developed programs to expand health insurance coverage or to improve access to care for the uninsured. These experiments provide a valuable laboratory to learn about approaches that might relieve a national system straining under the weight of 33 million people without health insurance, besieged hospital emergency rooms, soaring medical care costs and a myriad of other problems.

One approach employed by states has been to "tax" insured hospital patients to pay for those patients who lack health insurance and cannot pay their bills. Private health insurers agree to subsidize this uncompensated care in exchange for hospitals allowing the state to regulate their prices. This approach has dramatically relieved the economic difficulties of many urban hospitals and enabled more people to obtain hospital-based services. But employers who provide health insurance complain that they are paying for those who don't. And since only the hospitals — not the physicians — are reimbursed, many physicians are reluctant to continue providing free care as the number of uninsured patients mounts. A related problem is that uninsured patients come to hospitals for routine outpatient care that could be provided in much less costly settings.

A second approach has been to expand Medicaid programs, which provide a federal match for state funds used for hospital, physician, and other medical services for poor women, children, the disabled and the elderly. This approach is politically attractive because it targets "deserving" populations, and it is safe because the Medicaid bureaucracy is a known entity. Yet there are wide gaps in Medicaid coverage, and the persistently low reimbursement rates in most states have resulted in a severe shortage of physicians willing to serve Medicaid clients.

A third category of experiments involves state subsidies for private insurance. Since three-fourths of the total uninsured population comes from workers and their families, states have tried to make private insurance more affordable for both small employers and workers. One way has been to help small companies join together for insurance purposes, since rates can drop by as much as 20 percent as the size of the "risk pool" increases. Another idea has been to target family dependents not covered by health insurance plans. Some of these efforts have worked well, but it will be difficult to broaden them to cover large numbers of the uninsured without a great deal more money.

Finally, since 1974 Hawaii has required employers to offer a basic package of health benefits at a controlled price and to purchase health insurance for all employees. Mandated health insurance forces small employers to "pay their share" and avoids tapping state treasuries, but remains controversial because of the potential impacts on small businesses and on employment itself.

None of these four general approaches has emerged as the consensus candidate for a national plan. All have advantages and drawbacks. But the states are far ahead of the federal government in moving beyond rhetoric to see what really works. Their efforts deserve study — and support. Federal agencies should provide funding and technical assistance to help states evaluate these efforts and to experiment more broadly.

Once again, the states are our best laboratory. Some are engaging the health care issue with courage and vision, and serving their own citizens admirably while offering the nation a wealth of valuable experience.

July 28, 1991

Molly Joel Coye *is director of the Department of Health Services for California. This article is adapted from a longer version in the Summer 1991 edition of* Issues in Science and Technology.

★ ★ ★

Our Disabled View of Disability

Alvin R. Tarlov

When athlete Bo Jackson or singer Gloria Estefan suffers a disabling injury, the nation follows their rehabilitation anxiously. Why, then, are we doing such a poor job of helping 35 million other Americans with disabling conditions?

That's one of every seven Americans, making disability a problem that affects every neighborhood and family. Over a 75-year life span, the average American spends *13 years* with impairments that interfere with normal activities. The cost to the nation in medical treatment, income support to replace lost earnings, and rehabilitation services is more than $170 billion annually.

That is an immense toll. For the most part, it also is a preventable one. I chaired a committee of the Institute of Medicine of the National Academy of Sciences that recently formulated a national agenda for preventing disability. We found that millions of disabilities could be prevented or made far less disabling. The problem is not primarily medical, although improved medical care is required. Rather it is our collective complacency about allowing some conditions to progress needlessly to disabilities.

Disability is the gap between a person's capability and the demands or expectations of society. For too long we have viewed the former with resignation and the latter as beyond control. As a humane society that celebrates individualism in human beings and seeks to avoid suffering, it is a view we must change.

Most conditions that lead to disability are preventable. They include premature birth, malnutrition, social and educational deprivation in childhood, automobile crashes and other injuries, violence, alcohol and other drug abuse, and chronic diseases. When these conditions do occur, the resultant physical or mental impairment often can be reversed or minimized by prompt medical care and rehabilitative services. Even if the impairment becomes a permanent functional limitation, it need not prevent someone from working and participating fully in family and social life. But training, assistive devices, transportation, housing and adaptable workplaces are essential.

Consider stroke, which often leads to disability. Among the causes of stroke are lack of physical fitness, overeating, stress and failure to treat hypertension. All of these factors can be ameliorated, as can many causes of heart disease, respiratory problems, hearing loss and other disabling conditions.

Or think about the 1.3 million people who sustain head injuries each year. Many are teenage males involved in car crashes, falls or violence. For many of them, emergency care is too slow, rehabilitation services are inadequate, and long-term follow-up assistance is lacking to help the young men re-enter a full life.

Physicians and allied health professionals alone cannot solve these problems, although their roles should be expanded. To prevent so many Americans from slipping into dependency, the rest of society, including lawmakers, highway designers, employers, social workers, neighborhood organizations and others, also must get involved.

Insurance policies, for those having them, generally cover rehabilitation services only in acute-care hospitals, and only for the length of the hospital stay or for a month or two afterward. Coverage ends when hospitalization is completed and the condition has stabilized — just when post-acute rehabilitation should begin for many traumatic injuries and chronic conditions.

An elderly person who needs a lifting device to get into a car generally cannot get one through Medicare, which deems this and many other kinds of medical equipment as "convenience items." Incontinence pads are "non-reusable supplies." For want of a ride or out of fear of incontinence, some elderly people lose their mobility.

The Americans With Disabilities Act passed last year was a historic accomplishment, prohibiting discrimination against people with disabilities in hiring, transportation and other activities. But the greater challenge remains. We must *prevent* the injuries, birth defects and other conditions that often lead to disabilities, and deter existing conditions from progressing to the point of disability. When impairments do occur, we must strive to overcome them — not only with better medical care but with social services, job training, and much more to promote independence and improve the quality of life.

As many as two of every three disabilities could be prevented through disease and injury control, better health services and other initiatives. Millions of Americans now are disabled simply because we lack the will for prevention.

June 2, 1991

Alvin R. Tarlov *is senior scientist at The Health Institute of the New England Medical Center and a professor of public health at Harvard University.*

* * *

The Deadly Threat of Emerging Infections

Joshua Lederberg and Robert E. Shope

Most Americans wrongly view acute infectious diseases as a problem of the past. Actually, the danger posed by infectious diseases remains very real. We are likely to face deadly threats from new diseases and from the re-emergence of old ones. Our public health system needs to do much more to protect against these threats.

The AIDS epidemic illustrates how quickly an unknown disease can emerge to wreak havoc. Within a few years, it became one of the most urgent problems of our time.

In little more than a decade, meanwhile, the number of cases of Lyme disease reported each year to the Centers for Disease Control (CDC) has grown from a handful to more than 9,000.

Public health officials also are battling new virulent and drug-resistant strains of tuberculosis and malaria, and new forms of streptococcal bacteria.

We chaired a committee of infectious disease experts convened by the Institute of Medicine of the National Academy of Sciences. Its members agreed unanimously in a recent report that the threat from these and other infectious diseases is growing, posing serious challenges to our citizens.

Many factors account for this. Modern forms of transportation help microbes move quickly around the globe. Suburban development and reforestation of farmland in the United States and expanding human populations and deforestation elsewhere bring people into closer proximity with pests and other animals that transmit the diseases. People around the world who suffer from the diseases often have primitive or nonexistent medical attention. Most important, many mosquitoes have developed resistance to pesticides, and the disease organisms themselves have become resistant to drugs.

To protect ourselves, we must do several things. First, we need a much better surveillance system to detect unusual clusters of disease and track outbreaks, both in the United States and abroad. Improved surveillance also can help us understand why a disease has emerged. Without it, we are fighting blind. Our obliviousness to an increase in tuberculosis in New York City in the late 1970s, for instance, helped that disease re-emerge into a major menace today.

The CDC or some other federal coordinating body must expand current surveillance efforts. For example, the CDC needs the resources to do a better job of monitoring infections acquired in hospitals. One in 20 patients in U.S. hospitals, or 2 million people a year, pick up such infections, and 20,000 die from them. The U.S. Public Health Service should develop a more comprehensive data base that enables doctors to share relevant information quickly.

Second, we need a new arsenal of drugs, vaccines and pesticides. The United States should stockpile existing vaccines, increase its "surge capacity" to expand vaccine production quickly, and provide new incentives for manufacturers to undertake innovative approaches to vaccine development. We also need pesticides that are more effective and environmentally acceptable to control the insects and other creatures involved in public health emergencies.

A third need is to raise the level of informed, professional concern. Primary care physicians are the first line of defense against infectious diseases, but many of them are ill-informed about exotic and emergent diseases. We need more alertness, more research, more training in public health and related disciplines, and more education to help ordinary people protect themselves.

The medical community, political officials and ordinary Americans need a wake-up call. Some of the efforts we recommend require additional spending — an unpopular theme today. Yet these costs are far less than the potential consequences of inaction. Every dollar spent on tuberculosis prevention and control in the United States results in a savings of at least $3 to $4 in treatment. Influenza pandemics in 1957 and 1968 produced an economic burden of $26.8 billion. The AIDS epidemic is likely to cost much more.

Although we do not know where the next microbe or virus will appear, or how it will make us sick, we know that new outbreaks are certain. Unless we become more vigilant, some of these outbreaks could become new deadly epidemics, outdoing even today's AIDS crisis or the influenza pandemic that killed 20 million people worldwide after World War I.

November 22, 1992

***Joshua Lederberg**, a Nobel laureate, is University Professor and Sackler Foundation Scholar at Rockefeller University in New York. **Robert E. Shope** is professor of epidemiology and director of the Yale Arbovirus Research Unit at Yale University School of Medicine.*

★ ★ ★

Taking Women's Health Problems Seriously

Mary Lake Polan

Female baby boomers are entering the doorway of middle age. One thing that many of them can look forward to is fibroid uterine tumors, the major cause of hysterectomy in this country. Many others already have endometriosis, a disease that can cause crippling pain and infertility.

Both ailments are so common as to be considered "garden variety" problems. Yet they cause American women untold hours of misery. The medical causes remain obscure; cures are elusive.

When it comes to health problems like these involving obstetrics and gynecology, our lack of knowledge is enormous. Why are babies born prematurely? Why have ectopic pregnancies, those outside the uterus, increased every year since 1970, with a fatality rate of 42 per 1,000 cases? Why hasn't our understanding of sexually transmitted diseases like herpes, genital warts and AIDS caught up with the sexual revolution?

Why? Because our effort to find answers has been half-hearted and inadequate. I served recently on a committee of the Institute of Medicine of the National Academy of Sciences that examined research at academic departments of obstetrics and gynecology across the country. We found that only a few are providing a research environment as vibrant as one finds for AIDS, heart disease, Alzheimer's and other health problems. Women's health issues are being consistently devalued.

Faculty members in OB/GYN departments are carrying out significantly less research than those in other medical departments. Their research proposals are much less likely to receive funding from the National Institutes of Health.

This lack of scientific inquiry has a direct impact on clinicians such as the obstetrician who monitors a young pregnant woman's blood pressure and discovers that it is elevated, signaling the onset of pre-eclampsia. The obstetrician worries about her condition and that of the fetus. Is the fetus growing normally? Is there adequate fluid surrounding the fetus in the uterus?

Researchers could help that obstetrician learn why preeclampsia occurs and how to help both mother and child. Instead, too many fragile, growth-retarded babies continue to be born from the condition.

Another example is uterine fibroids. Also called a myoma or leiomyoma, a fibroid is a single muscle cell that grows and divides to form a benign muscle tumor in the middle of the uterus. It is a common occurrence. Fibroids grow slowly and unnoticed until they start causing pain or, more seriously, heavy and frequent bleeding. Sometimes they become so large that they resemble a pregnancy, causing bowel and bladder symptoms. Although not malignant, fibroids may cause so much pain and bleeding that a hysterectomy is necessary.

Scientists do not know why that single muscle cell begins to grow into a fibroid, much less what to do about it. The same is true for a multitude of other OB/GYN problems which, beyond their human toll, cost billions of dollars a year.

Research to answer these questions requires qualified researchers. Now there's a shortage. Those in positions of leadership need to provide more training and support to encourage bright young OB-GYNs to pursue research careers.

Women represent nearly half of all OB/GYN residents and face special obstacles in choosing the laboratory over clinical practice. The problems include coping with pregnancy and childbirth during crucial early faculty years, isolation from traditional support networks and a dearth of female role models. Providing mentors, flexible work arrangements or extended time to gain tenure could ease these pressures. Also needed is more government-sponsored support of OB/GYN research. Because of social and ethical associations with abortion and other controversies, reproductive health issues have become political pariahs and been neglected in NIH research budgets. The women of this country have been the losers. A greater commitment is needed not only from government but also from private industry.

Our committee prepared a list of research initiatives that would go a long way toward correcting the current sorry state of affairs. The list covers menopause, ovarian and uterine cancer, infertility, genetic development, child bearing, contraception and other issues that collectively affect almost every woman in the United States. One would think that's a large enough group of people for these health problems to start getting the attention they deserve.

June 28, 1992

***Mary Lake Polan** is professor and chair, department of obstetrics and gynecology, Stanford University Medical School.*

★ ★ ★

Pregnant Women, Newborns and AIDS

Marie C. McCormick

By the end of this year, health experts expect there to be a new entry among the five leading causes of death for women of reproductive age in the United States: AIDS. And because women can transmit the disease to their newborns, AIDS deaths also are increasing among young children.

More than 14,000 cases of AIDS among women were reported as of October in this country. The large majority of women with AIDS are between the ages of 15 and 44, which helps explain the increase in the number of children born infected with HIV, the virus that causes AIDS.

Society must act decisively to prevent additional suffering from AIDS in these new populations. But one thing we should *not* do, and which some health experts have suggested, is mandate routine screening of all pregnant women and newborns for HIV.

Not that screening is bad; it can be extraordinarily valuable in identifying patients who need treatment. Pregnant women infected with HIV who have severely depressed immune systems can benefit from current AIDS therapies. Women who learn they are infected also can make more informed reproductive choices.

So the problem is not "screening" but "routine." I recently chaired a committee of the Institute of Medicine of the National Academy of Sciences that examined HIV screening of pregnant women and newborns. We concluded that the HIV test is unlike most routine medical tests because a positive result can carry such profound psychological and social consequences. Individuals should take it only by choice.

Routine does not mean mandatory. In theory, women would retain the right to refuse a routine HIV test. Yet, in reality, many women might not know they can refuse the test. It is essential that HIV screening remain voluntary and that women be tested only after providing written informed consent.

The reason for this caution is that many women now infected with HIV face discrimination in employment, housing, access to health care and insurance, as well as in obtaining reproductive health services. It is unethical to mandate screening if needed counseling and follow-up services cannot be ensured for those who test positive. In many instances, such a guarantee is impossible.

With confidentiality safeguards in place, voluntary HIV testing should be offered to all pregnant women in areas where HIV is highly prevalent. Women who have not been tested prenatally should be offered testing at the time of delivery, or soon thereafter, with appropriate counseling before and after the test.

Not every state will choose to provide this service. Rates of HIV infection among women vary widely across the United States, and some states — particularly those with few cases of infection — may decide to spend their resources on other health services. State health authorities are in the best position to determine whether an HIV screening program is the best way to spend limited funds. Those who do decide to offer HIV screening must be committed to developing adequate health and social services for HIV-infected women and their children.

The case for routine HIV screening is even less compelling for newborns. Although an infant's own blood is examined, the test reveals only whether the child's mother is infected with the virus. If so, there is roughly a one-in-three chance the infant will be infected. Treating all infants born to HIV-infected mothers would result in all uninfected infants being exposed to toxic therapy without deriving any medical benefit. Also, newborn screening is tantamount to involuntary maternal screening, which is ethically unacceptable.

What *is* acceptable, and already being done across the country, is something quite different — anonymous HIV screening of newborns by public health researchers tracking the course of the epidemic.

The calculus of routine screening could change. The argument for it becomes more convincing as the certainty of the benefits to those who test positive — in the form of treatments or a cure — increases. For example, if a definitive test for newborn HIV infection and safe, effective treatment for infected infants were available, routine screening of newborns might be warranted. For the time being, however, the only routine thing about HIV screening should be that it remain anonymous for newborns and strictly voluntary for pregnant women.

February 17, 1991

Marie C. McCormick *is associate professor of pediatrics at Harvard Medical School.*

5

DIET AND NUTRITION

Weight Control: What Really Works

Judith S. Stern

The next time you see a commercial in which celebrities extol a diet program, be skeptical as to whether their personal success proves the effectiveness of the program for others.

A panel of the National Institutes of Health reported earlier this year that commercial diet programs have extremely low rates of long-term success. Ninety percent or more of the people who enroll in them regain all or most of their hard-lost pounds within five years.

That may sound like bad news for the millions of people in the United States — more than a third of the women and nearly a quarter of the men — who are trying to lose weight. An even larger number of Americans seek to maintain their weight. If commercial diet programs have such limited success rates, what possibility does anyone have of succeeding?

People would have a greater chance if they paid more attention to scientific findings and less to advertising claims. Many Americans worry needlessly about their weight, persuaded incorrectly that they are too fat. They even may harm themselves if they repeatedly lose weight and then gain it back. But those who really need to take off some pounds should not become discouraged.

The fact is that some people *do* lose weight and keep it off. Why are they successful while others keep yo-yoing through weight changes?

Much of the answer can be found in what we have begun to learn about successful "maintainers," data discussed recently at a meeting of the Food and Nutrition Board of the National Academy of Sciences' Institute of Medicine. I was part of a research project that studied three groups of women. The first group included women who never had a weight problem. Women in the second group lost weight and kept it off for at least two years. The third group was comprised of women who lost weight only to regain it.

The women in the first two groups shared several characteristics. They generally exercised regularly and ate foods lower in fat and sugar. Both of these findings make sense. Many research studies demonstrate that exercise helps people maintain lower weights. Experimental animals that exercise are much more likely than their "couch potato" counterparts to choose a low-fat diet. Calories from fat, meanwhile, are more likely to make people gain weight than calories from carbohydrates. In animals, a combination of fat and sugar is more fattening than fat and sugar eaten separately.

Exercising regularly and cutting back on fat and sugar are not the only successful weight control techniques that we identified in controlled studies. Women who never were overweight or who kept excess weight off successfully weighed themselves regularly. They were more likely to have good problem-solving skills and to confront difficulties directly rather than avoiding them. In addition, these maintainers usually had good social support networks with friends and family available to help them deal with everyday problems.

Emerging research also reveals interesting differences in how unsuccessful "regainers" and successful maintainers pursue their diets. Those who regain weight are more likely to follow doctor's orders and structured diet plans. In contrast, maintainers tend to follow a diet program they devise themselves to suit their own lifestyles. They take control and are "in the driver's seat."

Scientists still do not know the basic cause of many obesities. Even though Americans are obsessed with dieting, much of the research needed to provide more rational treatments has not yet been performed. Nonetheless, completed studies do suggest that women who seek to lose weight should choose an individual program tailored to their own needs, modify their food choices and be physically active. To keep the weight off, they also must learn effective problem-solving techniques. Admittedly, these skills are not easy to master.

One of the most encouraging things we have learned is that people often need not lose much weight to enjoy significant health benefits. Losing as little as 5 percent to 10 percent of body weight can bring about improvements in blood pressure, diabetes and blood lipids. In other words, for those who truly are overweight, making a change is well worth the effort. But the change needs to be based on science rather than on celebrity endorsements.

August 2, 1992

Judith S. Stern *is professor of nutrition and internal medicine at the University of California, Davis.*

* * *

Serving Up Nutrition Instead of Guilt

Edward N. Brandt Jr. and Paul R. Thomas

Trying to eat a healthful diet these days is no piece of cake. It's little wonder so many Americans don't eat as well as they should.

Many people do try to limit certain foods or to improve nutrition for their children. But anyone who has ever flinched at the bathroom scale or examined the wares at the local supermarket knows how demanding it can be to maintain a healthful diet.

If cars crashed hourly at an intersection, we'd redesign the intersection rather than blame all of the drivers. Yet we

Drawing by David Kiphuth
The Daily Gazette, Schenectady, N.Y.

tend to view the eating problems of countless Americans as signs of personal weakness rather than of something awry with the broader situation.

Throughout the United States, per capita consumption of fresh fruits and vegetables, breakfast cereals and low-fat milk has increased. Many consumers report using less salt and fat in their foods. However, people also are eating more high-fat cheeses and frozen desserts, fats and oils, snack foods and candy.

Perhaps consumers are being too hard on themselves in viewing such lapses as personal failures. The path to better nutrition is hindered by too many barriers, from fast foods loaded with saturated fat to incomplete labels on packaged foods. Surveys indicate that many people lack both the information and skills to eat better. What they need is not guilt but more help.

Thanks to continuing efforts by government, the food industry, health care providers and educators, it's easier than ever to eat well. The recent work of the federal government to develop better food labels, for example, could do a great deal to promote better nutrition. Yet it is possible to do more — much more. We served on a committee of the Institute of Medicine of the National Academy of Sciences that recently identified numerous ways to make things easier for eaters.

Supermarkets, for example, could provide nutritional information at points of purchase, such as with signs in the produce or dairy aisles, or with tags indicating that certain products are a good nutritional buy. Restaurants could provide nutritional analyses of their menu items.

Educators could help Americans make sense of this information, teaching pre-schoolers good eating habits along with their A-B-Cs and providing their older brothers and sisters with nutrition education throughout their schooling.

Company cafeterias, television shows and others could promote nutrition education more informally. Just as television producers have reduced the gratuitous use of alcohol and drugs on many programs, so might characters begin eating healthier foods. On the show "Cheers," for example, why not have Norm eat something healthier than beer nuts? Informal efforts are especially important for changing eating patterns among less educated and poorer Americans, who generally have worse diets.

Then there is government, which has tremendous leverage to encourage better nutrition through the School Lunch Program, farm subsidies and a host of other initiatives. At the very least, facilities such as public hospitals and government agency headquarters should provide good foods in their own cafeterias and kitchens.

Consumers who shop carefully in supermarkets generally can put together a healthful diet, and food producers have improved the content of many items by cutting back on the fat, salt and sugar. Yet consumers also need access to nutritious foods elsewhere. For many people, "dinner" means a stop at a fast-food restaurant or a microwaved frozen entree rather than a home-cooked meal. Instead of gorging on fat, they should be able to eat a healthful meal with a minimum of bother.

Consumers need help not only from restaurants and food producers, but from government, health care providers and everyone else involved with food in our country. Of course, we still must take responsibility for what we eat. But it should not be so difficult or confusing to eat wisely.

As matters now stand, our national diet is too high in guilt and too low in practical information and healthful choices. Rather than blaming people for failing to overcome the situation, it's time to cook up something better.

July 14, 1991

***Edward N. Brandt Jr.**, executive dean of the Health Sciences Center at the University of Oklahoma, Oklahoma City, chaired a committee of the Institute of Medicine of the National Academy of Sciences that examined the nation's diet and health. **Paul R. Thomas** was the committee's study director.*

* * *

The Foods in Our Future

Sanford A. Miller

Dieting baseball managers and talk show hosts get the publicity, but a quiet revolution now gaining momentum within the scientific world could change the way people lose weight — and much more. It could lower the risk of heart attacks, alleviate hunger and transform agriculture.

Breakthroughs in molecular biology and other disciplines are about to change the way we Americans produce, eat and think about food. These changes could yield great benefits. They also pose challenging social and scientific questions that warrant much more attention.

Scientists are beginning to use genetic engineering and

other tools to design foods with incredible precision. In the past, agricultural scientists developed more bountiful crops and healthier animals through traditional breeding — a slow, almost random process. But now researchers are speeding up and refining this process, like carpenters using power tools after centuries of working by hand.

The fruits — and vegetables — of their efforts are emerging in the form of fat substitutes, substitute lobster meat, crops with higher nutritive value and animals more resistant to disease. Other exciting foods with new shapes, colors and flavors are on the way. Before long, consumers could see foods made from fungi and other non-traditional materials — tasting like meat but more nutritious. Foods will have their genes reshuffled to produce desirable characteristics, such as fresh tomatoes in winter that actually taste like tomatoes right off the vine, and grains that thrive in drought-ridden areas of Africa.

Changes like these, when combined with progress in nutrition and other scientific fields, could improve the quality of life significantly. Yet, like any advance, they may cause harm if not implemented rationally. A recent conference of the Institute of Medicine of the National Academy of Sciences considered some of these problems.

The public outcry over the introduction of milk derived from animals treated with bovine somatotrophin (BST), and about using ionizing radiation to process food, shows that new food technologies can arouse considerable public fear. While most scientists agree that these technologies are safe, the controversy about them illustrates that any advance must be as acceptable socially and economically as it is scientifically.

One specific problem that could arise with new foods is the possibility of varied products being made from a diminishing number of raw materials, such as from soy beans. People need to eat a wide array of foods. Yet some companies may find it cheaper and technologically easier in the future to create look-alike foods from a few raw materials. For instance, they might create substitutes for beef, poultry or fish. Consumers might suffer nutritional imbalances if the foods were not carefully enriched.

With more efficient technologies, companies also may use fewer processing plants. This should yield cost savings and make the businesses more competitive, which helps our economy, but it could cost some workers their jobs. Increased consolidation also means that problems at any plant — such as an outbreak of food contamination — may affect more consumers.

The techniques used within the plants pose still other problems. Fermentation methods, for example, are likely to be used increasingly to produce everything from abalone substitutes to vitamins. Millions of consumers will enjoy these products. But fermentation also will produce waste materials that typically contain a large number of living microorganisms and organic materials. Methods will be needed to dispose of these wastes safely, perhaps by sterilizing them and converting them into animal feed or fertilizer.

Then there are international concerns. New food technologies are likely to be alluring to developing countries where hunger is common. Providing nutritious and hardy new foods for Ethiopia, for example, would be a major scientific accomplishment. Yet poor countries also will need help in regulating these advances to prevent abuses and to assure the safety and quality of their food.

As one involved in evaluating new foods, I think the benefits are likely to vastly outweigh these and other potential problems. Both the scientific community and public officials are considering these questions now to ensure that our food supply remains healthy, safe and economical in the years ahead. And if we plan carefully, the foods emerging from our laboratories should bring us everything from better nutrition to lower grocery bills. If we don't, however, tomorrow's menu may not be to everyone's liking.

March 31, 1991

Sanford A. Miller *is professor and dean of the Graduate School of Biomedical Sciences at the University of Texas Health Science Center at San Antonio.*

★ ★ ★

Improving the Safety of Seafood

John Liston

Few feelings are worse than eating some oysters, clams or other seafood and then, a few hours later, becoming sick with nausea and diarrhea.

Fortunately, most seafood sold to the public is wholesome and unlikely to cause illness if properly prepared. But Americans are eating a lot more seafood than they did just a decade ago, so the importance of ensuring a safe supply has intensified.

I chaired a committee of the Institute of Medicine of the National Academy of Sciences that recently examined seafood safety. We found that seafood generally deserves its reputation as a healthy food. The number of reported cases of seafood-borne illness has remained fairly low despite increased consumption. Yet seafood safety can be strengthened.

This is best done by preventing sewage, hazardous chemicals and other contaminants from entering the waters where shellfish and finfish are caught. The cleaner the water, the safer the fish are likely to be. Unfortunately, cleaning our fishing waters will take years and even then will not put an end to problems due to naturally occurring bacteria or marine toxins. So other actions are needed now.

What's *not* needed is a lot of new inspectors sniffing and probing individual fish before they are sold. This would be terribly costly and essentially worthless for detecting or controlling health risks.

The biggest single risk associated with eating seafood is from raw oysters, clams, mussels and other "bivalve molluscs," which may cause hepatitis and other problems if contaminated or mishandled. Raw shellfish is especially risky for people with liver disease or weakened immune systems. Health officials should do more to warn the public about this danger, and consumers always should cook seafood thoroughly.

A quite different kind of problem, and one that cannot be

eased by adequate cooking, is fish contaminated with natural toxins. For example, fish caught near reefs in tropical areas may contain a toxin that causes ciguatera, a particularly nasty kind of food poisoning. The best way to prevent ciguatera is by banning certain fish species or fish caught in suspect areas, and by developing tests to identify toxic fish.

Then there are toxic chemicals such as PCBs or dioxin, which can accumulate in aquatic animals. Evaluating chemical health risks often is difficult since the effects — such as an increase in cancer — may be slow to appear and hard to attribute to a single cause. But the threat can be reduced by restricting harvests from polluted waters.

Supervision is done most effectively on the state and local level since problems vary so widely from place to place. While consumers in Puerto Rico worry about ciguatera, for example, those on the mainland may be more concerned about a milder form of poisoning in fresh and frozen species like tuna or bluefish. In Alaska, botulism can occur in some traditional fermented seafood.

Consumers are now protected, although inadequately, by an intricate system of federal and state programs. Inspectors at all levels focus too much on the "finished products" — the fish in the store — rather than on the raw materials in the water. The most important federal role should be to provide guidance and technical assistance and, since imported seafood accounts for more than half of U.S. consumption, to monitor the inspection procedures of foreign countries.

Local officials, meanwhile, need better enforcement measures to prevent "bootlegged" shellfish from hazardous areas from reaching the market. They also might restrict harvests of some shellfish during warmer weather, when the threat of food poisoning is greatest.

Scientists and other technical experts can help, too, by providing more effective methods of testing whether waters or individual fish are contaminated. A simple test to screen fish for ciguatera has been developed but needs more examination. Improved tests for viruses could help protect shellfish waters. Better technology and education are needed to improve safety for the millions of Americans who fish for sport or daily subsistence.

Despite these concerns, seafood remains one of the safest and healthiest foods consumers can eat, whether in a fresh salmon steak, trout almondine or a tuna sandwich. But with more and more Americans eating seafood, safety now needs to become as inseparable from their meals as the lemon slices and tartar sauce.

January 27, 1991

John Liston *is professor emeritus of the Institute for Food Science and Technology at the University of Washington in Seattle.*

* * *

Fighting Trim, Fighting Smart

Robert O. Nesheim

An old adage states that an army travels on its stomach. But Army standards say too much stomach is unacceptable; soldiers who get fat risk being dismissed.

Requiring soldiers to be in "fighting trim," as currently defined, is of increasingly dubious value in an age when success on the battlefield depends as much on programming computers as on charging up a hill. Dismissing a brilliant computer programmer because he or she is somewhat overweight does not make any more sense for the military than it would in academic or business situations.

Not that the military should disregard the physical condition of its troops. But it does need to find a better match between its weight standards and people's ability to perform military tasks.

I recently chaired a committee of the Institute of Medicine of the National Academy of Sciences that examined the Army's standards and found them in need of updating. Good people may be forced out of the military or deterred from joining because their physiques fail to measure up to an ideal of questionable relevance.

The Army recently revised its weight standards to ease a disparity that was especially unfair to women. Females had been allowed less variation over their prescribed standards than their male counterparts were allowed over their standards. Racial and ethnic groups also have diverse body types that may differ from norms established by years of studies of Caucasian males.

Making the standards more equitable among the diverse members of the military is commendable. But changes still are needed in the system as a whole.

According to Army regulations, a six-foot male recruit, 17 to 20 years old, should weigh no more than 200 pounds. A five-foot, five-inch woman of the same age should weigh no more than 141 pounds. If the recruit weighs more, the Army then determines the proportion of body fat through a number of different procedures. Once admitted to the military, personnel must meet even more stringent retention standards. Both recruits and service personnel also must meet fitness standards that include push-ups, sit-ups, running and similar exercises.

Seemingly sensible, these measures do not test directly what the Army really needs to know, which is whether someone can do the job. Smaller, lighter individuals who do well in running and physical fitness tests may perform poorly on tasks such as lifting and load-carrying. A moderately overweight person, by contrast, may be better at pitching tents for hurricane victims in Florida, carrying loads in the Persian Gulf or other tasks, not to mention jobs done at a desk. Our committee concluded there is no consistent relationship between body fat content and physically demanding jobs.

Lean body mass (muscle and bone) is a much better predictor than body fat of physical performance. The services should consider establishing minimum standards for lean body mass and also should develop specific standard tests that more accurately reflect military activities.

Currently the military standards primarily involve height and weight. As a result, healthy individuals who are very muscular, such as professional football players, may be rejected for being too heavy.

Does all this mean the military should not be concerned about excess weight or military appearance? Of course not. About 20 percent of the U.S. population is overweight and obese, and there is a proven correlation between obesity and diverse health problems. The armed services, which provide medical care for millions of employees and retirees, have a legitimate interest in minimizing these costly problems.

Furthermore, for purposes of public image and unit morale, the services have a right to expect a reasonably trim military appearance among those in uniform. But holding down health care costs and keeping soldiers looking sharp does not require height-weight standards as stringent as those now on the books.

Revising these standards would not open the door to poor discipline and ranks filled with potbellies. Rather, they would reflect modern understanding of body composition and physical performance. As the military faces severe budget cutbacks, it must not squander the years of training it has invested in valuable personnel, based on out-of date standards.

September 20, 1992

Robert O. Nesheim *is retired vice president for science and technology at Quaker Oats Co.*

6

TECHNOLOGY AND TRANSPORTATION

Getting Serious About Computer Security

David D. Clark

We Americans have been remarkably lucky. As far as we know, no one has systematically subverted our critical computing systems. Not yet.

There are signs our luck soon may run out. Thousands of computer "virus attacks" have been reported, money and information have been stolen successfully and lives have even been lost because of computer errors. A German computer club broke into NASA's computer. A student injected a "worm" into a nationwide computer system. Hackers have taken over TV satellite linkups. Patient information in a Michigan hospital computer was altered by a virus. A computer expert nearly defrauded the Pennsylvania Lottery of $15.2 million by pirating unclaimed computerized ticket numbers.

Some of the most serious problems have been unintentional. A year ago, for example, a software design error froze much of the country's long-distance network. Nonetheless, the nation has not yet suffered a truly catastrophic computer breakdown or security breach.

However, whether due to sabotage, poor design, insufficient quality control or an accident, the problem of computer security is very real — and growing. The advent of widespread computer networking and increasing computer literacy among the public has brought us to the point where we must all begin taking computer security seriously or suffer the nearly inevitable consequences.

As matters now stand, far too many of the nation's computer users are like people in a small town who leave their houses and cars unlocked because they feel secure. Yet, with every day that computer systems become more prevalent and interconnected, this bucolic view becomes more outdated — and dangerous. Hackers receive the most attention, but they probably are less of a threat than disgruntled employees, terrorists and others who could wreak havoc on inadequately protected computers.

I chaired an expert committee of the National Research Council that recently studied this problem, and we identified some specific actions to improve computer security, both now and over the next decade:

- Companies and organizations must implement better policies to protect electronic access, physical security, networking and the like.
- Where appropriate, organizations should form computer emergency response teams to deal with security violations.
- A systematic effort should be undertaken to gather data on computer crimes.
- Universities should provide training in security engineering.
- U.S. computer manufacturers and software designers must improve the security features of their products to remain competitive with foreign firms.

One of the most important needs is a clear articulation of what constitutes basic computer security — a set of security principles comparable in acceptance and scope to building codes or to the "Generally Accepted Accounting Principles" that guide the accounting industry. These requirements might build on criteria already established for the defense department, ensuring that a minimal level of security exists for all computer users. The goal would be to "raise the security floor" and provide a safer computing environment for everyone.

Raising everyone to a minimum level of security is important because often it takes only one weak link in a computer network to permit access to the entire system. The current situation is like a polluting factory in a densely populated area; one person's laxness can affect many others.

Accomplishing these goals will require sustained collective action by the group at risk: the users of computers. Companies should join to establish a private, not-for-profit organization to focus attention and provide research on security issues. No such body now exists and one is sorely needed, preferably outside the government. While the government does have a key role to play, a private organization would be best positioned to direct this effort.

These measures will take time. But there are several things everyone who uses a computer can do right now. People can rename their password to something unpronounceable, and change it regularly, so others cannot gain access to their files. They can avoid software that has not been checked for viruses, and turn off their systems when not in use.

Most important, everyone must wake up and recognize the threat. A thief today can steal more with a computer than with a gun; a terrorist tomorrow can do more damage with a keyboard than with a bomb. Our luck, and our computer files, could change in an instant.

January 6, 1991

David Clark *is a senior research scientist at the Massachusetts Institute of Technology.*

* * *

Preventing Oil Spills Here at Home

Henry S. Marcus

The Persian Gulf conflict has once again focused our attention on oil spills. The terrible spills off Saudi Arabia are

Drawing by Pat Bagley
The Salt Lake Tribune, Utah

vastly different in origin and size from those that have occurred near our own coastlines. Yet they should spur us to renew our efforts to prevent oil spills here at home. One of the best ways to do this is by improving the design and operation of the tankers that carry oil to our shores.

The need for this oil is growing. Energy experts expect imports of crude oil and petroleum products to increase by up to 50 percent within a decade. Most of this additional oil will arrive by sea. Less than one in every 50,000 gallons of the oil moving through U.S. waters is spilled, but that adds up to an average of 9,000 tons annually — more than enough to cause serious environmental damage.

Last year, in the wake of the *Exxon Valdez* disaster, Congress mandated that all new tankers traveling in U.S. waters be built with double hulls. Yet double-hulled ships alone cannot protect our coastlines from disaster.

I chaired a committee of the National Research Council that has just released a study of 17 possible tanker designs. We found that the double-hull approach has many advantages over traditional single-hull designs, especially in a low-speed collision or grounding. A severe accident such as occurred with the *Exxon Valdez*, however, probably would puncture both hulls and spill oil. The double-hull design also poses some potential safety problems of its own, such as difficulty in inspecting the large void space between the hulls.

These concerns are manageable; overall, we found no design superior to double hulls in all accident scenarios. But double hulls are not a panacea; all designs perform better in some situations and worse in others. One design that deserves more consideration, for example, has double sides and an oil-tight deck across its middle to divide cargo tanks into upper and lower chambers. In theory, such a vessel would spill less oil in a severe accident — although it, too, has drawbacks.

This and other alternatives look good on paper, but implementing them raises other concerns. When automobile manufacturers consider the feasibility of safety options, they crash vehicles into a wall and do other tests to see which options work best. Yet the maritime industry has done relatively few safety experiments, largely because it has not been expected to design craft to withstand accidents. As a result, it lacks the database or even the criteria to evaluate how well a vessel will survive an accident. No one really knows how double hulls and other designs will perform. This is unacceptable.

It also is difficult to assess the costs and benefits of different options. Our committee's best estimate was that double hulls could cut by half the oil now spilled in U.S. waters as a result of collisions and groundings. They would add one or two cents to the cost of each gallon shipped, or $700 million per year when the Oil Pollution Act of 1990 is fully implemented.

More information is badly needed. But in the meantime, it is essential to consider ways of controlling pollution more effectively on existing vessels. The phase-out of single-hull tank vessels in U.S. waters will begin only in 1995 and then take 20 years to complete. Serious consideration should be given to requiring that all existing crude-oil tankers in U.S. waters promptly meet the latest International Maritime Organization provisions for pollution prevention for new tankers. Structural or operational changes also might be required of existing tankers, although the cost and effectiveness of these need to be weighed against other possible safety measures, such as increased crew training or improved traffic systems.

Vessel design is an essential aspect of preventing oil spills, and the law mandating double-hulled tankers is a useful step. More needs to be done, however, and many questions about tanker safety remain. The Persian Gulf spills, although in a different setting, remind us of the catastrophe that could strike our own shores again if we do not become more vigilant and get some answers soon.

March 3, 1991

***Henry Marcus** chairs the ocean systems management program at the Massachusetts Institute of Technology.*

★ ★ ★

Looking Beyond Potholes

Damian J. Kulash

The next time you lose a hubcap in a pothole or are stuck waiting for a road crew to repave the highway, think of Iraq. No, not the Iraq of Saddam Hussein, but the Iraq of nearly 4,000 years ago. The asphalt we use has its origins in ancient Iraq, or Babylonia. One reason our roads aren't better is that asphalt is still mired in too much ancient mystery.

With the winter pothole season approaching, we need to get smarter about this essential material. More than half of our nation's highways need everything from repaving to structural overhauls. The cost will be billions of dollars. Recent progress in asphalt chemistry makes it possible to build pavements that ride smoother and last years longer. We don't have to live with more potholes, more delays, and more spending.

We should look beyond our outdated technology and take

advantage of advances in materials science and diagnostic technology to steer into the 21st century.

Asphalt has been used for thousands of years. Surviving roads from Babylonian times still show good adhesion and toughness. When these roads were excavated early in this century, the asphalt mortar was so strong that it was hard to separate the bricks. The Sumerians and Assyrians also distinguished among many separate grades of asphalt, one of which was poured over the heads of convicted felons as a punishment.

Asphalt fell out of use later in history but re-emerged as a paving material in the 19th century. Since then, however, relatively little has been done to study it scientifically. Now we're paying the price for that neglect. Today's paving practices are based more on experience than on science; materials and methods are used because they worked before. Many of these tried-and-true approaches are more tried than true, unable to meet the growing demands on our highways.

Until recently, asphalt was thought of as a sticky gel binding together discrete lumps of material. But new findings show it to be more like strands of spaghetti swimming in sauce. Researchers working under a program of the National Research Council have generated a chemical model that sheds new light on the inner workings of asphalt. Using advanced chemical analysis techniques and studies of asphalt properties, they learned about a specific component that helps determine how asphalt performs.

The component, known as an amphoteric, is a combination of an acid and a base in the same molecule. The acidic and basic sites bind together and form the "spaghetti" that gives asphalt its structure. Although amphoterics usually constitute less than a quarter of the asphalt, they affect its overall quality profoundly.

This advance will allow highway engineers to specify the asphalt required for given climates and road conditions. States are already conducting tests to help them select the specific asphalt cements and aggregates they need. Refiners and manufacturers, in turn, will satisfy these requests by determining the chemical and physical properties of asphalt cement. The result will be materials that stand up better to heavy trucks, winter storms and millions of cars.

These advances could not come at a better time. The number of sources of crude oil used to make asphalt has risen to more than 100 nationwide. Since the oil from each source is chemically and physically different, engineers cannot control the final product adequately. They've been like chefs choosing from an ever-changing list of ingredients. The only way for them to avoid a lengthy and inexact process of trial and error is to understand what is happening chemically inside the asphalt and learn to pick the materials that work best for each purpose.

That's now possible, and we should begin improving our road infrastructure not just in small increments, but dramatically. Together with new insights from Europe and elsewhere about the construction of asphalt mixes, state and local highway agencies are experimenting with technologies to build more durable pavements.

Change will take time. Highway agencies and contractors need to shift specifications and tests. They also require new expertise and testing apparatus. With help from the Federal Highway Administration, they are beginning to get the training and laboratory capacity they need. But, for the sake of the nation's motorists, we need to push even harder. It's time we made our roads a lot smoother, tougher and more cost-effective.

November 24, 1991

Damian J. Kulash *is executive director of the Strategic Highway Research Program of the National Research Council.*

★ ★ ★

Getting Smart About 'Intelligent' Vehicles and Highways

Daniel Roos

You're late for work, so you hop in your car and immediately check your dashboard control for a detailed traffic update. You decide to take a new route, so you call up a map on your electronic console to guide you. You drive along until suddenly an alarm inside the car warns you of a truck about to veer into your fender. You steer to avoid it just in time.

Next you encounter a toll booth. No problem; you zip through it without stopping to pay. Instead, a code on your car is "read" by an electronic device. You'll receive the bill in the mail next week. Finally, as you pull into the parking lot, a beacon on your dashboard guides you to an empty spot.

If all this sounds preferable to sitting in endless traffic jams, it should. Technologies like these could not only reduce traffic delays but also cut fuel consumption, ease air pollution, lower freight costs and save lives.

A committee that I chaired for the National Research Council concluded recently that high-tech approaches to traffic management are increasingly feasible. Many of the ideas are not new, and components of them have been in use for many years. However, recently there has been a resurgence of interest in finding bold new solutions to our traffic woes. Many concepts have become practical with the spectacular evolution and declining costs of computers and telecommunications. Meanwhile it has become increasingly difficult to solve transportation problems by building new roads.

Computers have transformed our workplaces but had relatively little impact so far on our road and transit systems. By applying the tools of the Information Revolution, we might improve travel dramatically. Imagine, for example, that vehicles were equipped with sensors and adaptive cruise control technology enabling them to automatically maintain a constant distance from adjacent vehicles. This would make it possible for vehicles to travel safely only inches apart, like railroad cars. Each lane could thus carry far more vehicles, reducing the need for new roads. Drivers might even be free to read a book or enjoy the scenery during the trip.

Some new technologies might develop as entirely private systems, with firms providing services to subscribers. For example, a traveler information system could be organized like a private cable television or cellular telephone service. Alternatively, new systems might be operated publicly, as are nearly all existing traffic management services. The best approach may be a partnership in which the government provides some components while private companies provide others.

The recently passed surface transportation law will greatly expand federal research and development of "Intelligent Vehicle and Highway Systems," or IVHS, and many states are supporting similar efforts. It is becoming possible to visualize a transportation system built upon a new kind of national "information infrastructure," just as the construction of the interstate highway system transformed travel in an earlier era. A newly created, private, non-profit organization called IVHS America is helping to coordinate and gain support for these initiatives.

A danger is that research funds intended to develop genuinely new ideas may be diverted for more cautious efforts. Public agencies have a natural tendency to stick with familiar approaches and try to maintain control over traffic management. Turf concerns are a related problem; squabbles between highway and transit officials could choke innovations that cut across current boundaries. What's needed is a national program that pursues truly creative ideas, one in which government is organized to accommodate new technologies.

It also is essential that the public and private sectors work together closely. Automobile and electronics manufacturers and others in the private sector have to join in meeting this emerging challenge. Major markets are likely to emerge both in the United States and abroad for new kinds of vehicles, electronics and communications. American companies can ill afford to fall behind their counterparts in Europe and Japan, where a great deal of activity already is under way.

Transportation of the future could look vastly different — and better — than it does today. Instead of beating our breasts about our worsening traffic jams, we should use our heads to solve them.

January 5, 1992

Daniel Roos *is director of the Center for Technology, Policy and Industrial Development, and Japan Steel Industry Professor at the Massachusetts Institute of Technology.*

* * *

A High-Tech Cure for Traffic Jams?

Lawrence D. Dahms

With our highways and airports more crowded every year, Americans can look forward to increasing gridlock and frustration unless something is done to solve our transportation headaches.

One tantalizing solution that has been proposed is to build high-speed trains like those found in Japan and Europe. Traveling up to 200 miles per hour, these trains could whisk passengers between New York and Washington, Dallas and Houston, San Francisco and Los Angeles, Orlando and Tampa, or elsewhere. They'd give travelers more options while making airports and highways less congested.

At the request of the U.S Department of Transportation, a committee of the National Research Council, which I chaired, examined the feasibility not only of existing high-speed trains but of more futuristic possibilities, such as "maglev" trains powered by magnetic levitation.

High-speed trains similar to those in Japan and Europe are technologically feasible right now. Unfortunately they are costly. Proposals for building such systems in the United States have ranged between $10 million and $20 million per mile, depending on the location. The most likely market is intercity trips of approximately 150 to 500 miles, with high-speed trains competing principally with air travel for ridership.

Getting people to use high-speed trains in the United States will be much tougher than it has been in Europe or Japan. Why? In part because substantial intercity rail ridership already existed in Europe and Japan when high-speed trains were introduced. Also, competition from the air and auto modes is likely to be stiffer in the United States. We have more frequent flights, lower fares and cheaper gasoline costs. When these factors are combined with the high cost of building and operating high-speed rail systems, it is almost impossible to imagine how ticket revenues could cover the full costs of new train systems in our country.

Those who dream of riding fast trains in the United States must face the fact that subsidies will be required.

Whether such subsidies are justified is more a political than a technical question. If building a high-speed train can be shown to be better than expanding airports or adding freeways between certain major cities, then perhaps the subsidies could be paid with money saved from the airport or highway funds. Alas, our governments are not organized to make such tradeoffs. This problem was discussed during the recent debate over the surface transportation act. But as matters now stand, highway funds are held in trust for highways, airport funds for airports, and no one has the power to make a meaningful exception.

Our committee offered four main messages:

- Fast trains *can* be built and operated; we are not unduly limited by technology. High-speed trains are more feasible for the near term than maglev trains, which still are in a relatively early stage of development.
- These systems cannot be paid for solely with revenues

from the fare box. Subsidies are necessary if real progress is to be made.

• Subsidies may be justified in specific cases if it can be shown that a rail investment makes more sense than building new highways or expanding airports.

• Congress and state legislatures must write new laws empowering both the U.S. Department of Transportation and the states to consider fast trains as an integral part of the national transportation system. Only then can subsidies be directed to appropriate projects.

This last conclusion is the most critical. Congress is expected to invest in research to advance the promise of maglev trains as a faster and perhaps less costly alternative for the distant future. This initiative is encouraging but, even if the research is successful, we still have to figure out how to get airplane and train operators working together to select the best investment for a particular transportation corridor.

In Los Angeles, for instance, a new airport is planned in Palmdale. Further north, San Francisco International Airport hopes to expand. Would the alternative of building a high-speed rail system to connect these two cities make more sense? No governmental agency now is empowered to make such a decision. Isn't it time such tradeoffs were made possible?

December 15, 1991

Lawrence D. Dahms *is executive director of the Metropolitan Transportation Commission, Oakland, Calif.*

* * *

Crossing the Bridge to More Beautiful Journeys

Frederick Gottemoeller

Suppose you were making a commercial to convey the sense of a product reaching consumers from coast to coast. Which visual images would you choose?

When United Airlines faced that problem a few years ago, it chose the Brooklyn Bridge to represent the East Coast and the Golden Gate Bridge for the West Coast. That's not surprising. Bridges are among our most celebrated structures. Stretching across Tampa Bay or the Mississippi River, they can become symbols for entire regions.

Our thousands of "everyday" bridges are important, too. The bridges on Chicago's Dan Ryan Expressway, for example, are seen by millions of commuters weekly. Collectively, those bridges have a greater impact than any single public building on people's perception of that city.

As any traveler knows, however, many bridges are sadly nondescript. Carrying traffic but lacking grace, they are merely functional.

They could be much more. They should be works of civic art that enliven each day's travel, making all of our journeys more pleasant.

I was among a group of engineers and architects that recently examined bridge design for the National Research Council. We identified a number of examples where engineers are designing bridges that are not only beautiful but reasonable in cost.

Two changes are needed to make such bridges more routine. First, people must insist that their public agencies make good appearance an explicit goal of public works. Second, the engineering profession must improve its ability to respond to that challenge.

When citizens speak up, more attractive bridges often result. In Columbus, Ohio, public interest encouraged officials to organize an informal competition to obtain the best possible design for the new Broad Street Bridge. In Tennes-

Drawing by Douglas Edwards
The Baltimore Sun, Md.

see and California, continuing public support has led to a tradition of outstanding bridges. In Maryland, a new three-year program will improve the appearance of state bridges.

Many people think improved aesthetic impact derives from expensive "add-ons," such as an unusual color, ornamental features, or special materials like stone or brick. But in fact the greatest aesthetic impact is made by the structural components themselves — the cables, girders and piers. If these elements are well-shaped, the bridge will be attractive without added cost.

The Golden Gate Bridge, for example, owes its appeal to the graceful shape of its towers and cables, not to its reddish color. If the towers were ugly, painting them red would not make them attractive.

John Roebling, designer of the Brooklyn Bridge, and other notable engineers of the past and present have designed bridges that achieve structural excellence and outstanding appearance at a cost no greater than competing solutions. Their success proves that beauty need cost no more than mediocrity.

The dreariness of many bridges is not due to a lack of good intentions among designers. Most engineers do believe that concern for appearance should be an integral part of their work. But other matters often grab their attention. Engineers respond to the priorities of their clients — state highway departments, toll authorities and other public agencies that now give aesthetics too little consideration.

These agencies have the most influence to improve the situation. They set the standards, select the designers and pay for the results, and they should demand more. They would do so if they thought the public expected it. To put it another way, Americans should insist on getting full value for their tax money. They should demand not just bridges but beautiful bridges.

Engineers can meet that demand although, as a group, they have some problems to overcome. Bridge design is an art, one that integrates judgments based on science and mathematics with others based on aesthetics. Many engineers are more comfortable with mathematical formulas than with the imprecision of appearance. Their bridges suffer as a result.

So engineers must expand their skills. The key is to make clear to them that appearance is a criterion equal in importance to performance and cost. Doing so will encourage engineering schools to place greater emphasis on aesthetic concerns and lead engineers to develop their aesthetic abilities in everyday practice.

The engineering challenge of building bridges is not just to find the "least cost" solution. It is to bring forth elegance from utility. We should not be content with bridges that move only vehicles and people. They should move our spirits as well.

September 1, 1991

***Frederick Gottemoeller**, a consulting architect and engineer in Columbia, Md., specializes in the design of bridges.*

* * *

Launching Into a New Era in Space

Joseph G. Gavin Jr.

It once was one of our country's proudest enterprises. But then its facilities began to deteriorate and its vehicles failed to keep pace with models from abroad. Today, unless it starts building engines and vehicles that are more reliable and reasonably priced, it will fall even further behind competitors from Europe and Japan.

No, this is not another article about the automobile industry.

It is about a problem occurring high above Detroit and the rest of the country — in space. The United States, the nation that sent men to the moon, is falling behind in its ability to launch vehicles into orbit and beyond.

Both the European Arianespace group and the former Soviet Union have achieved higher launch rates with greater efficiency and lower costs than in the United States. Japan and other nations are making long-term commitments to large undertakings in space.

Our country, by contrast, continues to depend on outdated technology. The launch vehicles used by government and private industry are fragile, expensive and unreliable. Only one major advance in U.S. launch engines, the main engine of the space shuttle, has been made in 30 years. This was a superb engineering achievement when introduced in the 1970s but now operates at the very upper limits of its performance margins. Development of the more robust Space Transportation Main Engine should proceed immediately.

Our launch facilities in Florida and California are several decades old. They are customized to accommodate each vehicle, causing costly scheduling problems.

A committee of the National Research Council, which I chaired, concluded recently that this serious situation is getting worse. To meet future national needs and compete with other nations, the United States must reduce its launch costs by one-third to one-half. Action by the federal government also could enhance the competitiveness of our country's budding commercial launch industry.

The administration has proposed a National Launch System to help revive our "Earth-to-orbit" capability. Yet the program's top priority is developing a launch vehicle to carry very heavy payloads, those in the 135,000 pound class. Although heavy loads could be lifted to low Earth orbit, it is not completely clear what this vehicle would be used for. One possibility might be to resupply a space station. The Space Station Office, however, has no firm plans for this capability. The same is true as regards lifting heavy loads for space exploration.

So the need for a heavy payload vehicle is unclear, and developing it will be expensive and complex. Instead, we should focus on a different model with a clearer purpose and lower cost. A much smaller vehicle in the 20,000-pound payload class would have immediate commercial and national security applications. It also could be produced at a rate allowing individual units to have a reasonable cost. Building it first would provide good experience for tackling larger models, including an intermediate design for payloads of about 50,000 pounds, which eventually could replace the Air Force's workhorse Titan IV vehicle.

In so doing, we should learn from the success of the former Soviet Union, choosing low system cost and high reliability over exotic technology. The United States needs access to an array of engines and should evaluate all possibilities, regardless of their origin. For example, we should examine the Russian RD-170 rocket engine used in the *Energia* and *Zenit* launch vehicles.

The National Aeronautics and Space Administration (NASA) is developing an advanced solid rocket motor for use on the space shuttle and other launch vehicles. It should reconsider funding this program due to high technical and programmatic risks. With constrained resources, NASA should keep using its redesigned solid rocket motor, which has proven reliable since the *Challenger* accident. For future "strap-on" booster applications, liquid boosters should be considered because they have several advantages over solid boosters.

The United States also needs to strengthen its long-term

investment in the technology base that leads to new approaches. Without adequate research and development, our launch vehicles and facilities are certain to slip even further behind. Companies with satellites, scientists with experiments and others will turn elsewhere to launch their payloads into orbit. This trend, already alarming, must be reversed. Whether on the highway or up in space, we cannot compete using outdated vehicles or facilities.

May 17, 1992

Joseph G. Gavin Jr. *is former president of Grumman Corp.*

7

THE ECONOMY

Mobilizing for a U.S. Technology Strategy

Erich Bloch

Now that the parades for returning Gulf War soldiers are over, we must turn our national will toward winning a battle of even greater consequence, and one we have been losing — the battle for economic security.

Our country's security depends as much on its economic performance as on its military strength. But having recently completed six years overseeing many of our country's efforts in the scientific and technical arena, I have great concerns about whether we can compete unless we get serious about stopping the erosion of our human resource, technology and manufacturing bases.

Ten years ago the United States exported three to four times as many high-tech products as it imported. Today such imports and exports are roughly equal. Our positive trade balance in computers has been virtually wiped out, and our share of the world semiconductor market has tumbled from 57 percent to 35 percent. Our position is declining even in commercial aircraft manufacture, where U.S. dominance recently seemed as impregnable as it was in automobiles 25 years ago.

We have lost or are losing the battle for advanced ceramics, precision bearings, laser devices, fuel-efficient engines, optical information display and storage systems, and many other critical technologies.

Some see this trend as inevitable, saying the United States could not possibly maintain the technological supremacy it enjoyed after World War II. But the danger we face is not others catching up with us. That occurred a long time ago. The issue today is that other countries have surpassed and are continuing to surpass us in optical glasses, industrial instrumentation, electronic consumer products and their underlying technologies, and many other fields. We cannot do without a presence — or even be second — in the basic technology and manufacturing sectors of the 21st century without simultaneously foregoing our political leadership.

While industry has the primary responsibility for developing new technology and products and for earning market share, the policies and activities of the federal government have a major impact. Traditionally, Washington has supported technological development mainly through military and space programs, leading to early U.S. leadership in aerospace, computers and semiconductors. With the end of the Cold War and the emergence of the civilian sector as the main source of technological innovation and initial usage, this approach to technology policy is inadequate.

President Bush and Congress must acknowledge this new reality. Just as we crafted a national science policy a half-century ago — a decision that led to our becoming the unquestioned world leader in basic research and higher education — so should we develop a technology policy to reverse the decline of our civilian technology base. Otherwise, as foreign competitors aggressively pursue public-private initiatives, further erosion is virtually assured.

The U.S. government now spends more than 60 percent of its research and development budget on defense. Changing this to a 50/50 split would bring our civilian effort closer to that of our major foreign competitors.

Nearly a third of the government's R&D funds go to support federal laboratories, mostly for defense applications. Although often doing outstanding work, the labs are underutilized. Some of them should work more closely with industry to develop U.S. leadership in generic technologies of strategic importance, such as electronics, biotechnology or advanced materials.

Federal agencies should support collaborative development of generic technologies by industry and universities, and fund better mechanisms to transfer ideas from the laboratory to the marketplace. Several recent initiatives, such as the semiconductor industry's SEMATECH consortium, Engineering Research Centers, and new regional manufacturing technology centers, provide possible models for such partnerships. Collaboration among government, industry and universities can be pursued without undue worry about "picking winners and losers." Unsubstantiated concern about interfering in the marketplace must not become a rationale for doing nothing.

Technology policy also needs to support education, help upgrade the skills of existing workers, and provide tax credits and other financial incentives to promote research in industry.

We are at a turning point. Our science and technology policies are based on the realities of the 1950s, not those of the 21st century. They are obsolete and should be replaced by a strategy of collaboration among the sectors of our society, one that will stem the continuing loss of American technological leadership.

August 4, 1991

Erich Bloch *recently completed six years as director of the National Science Foundation.*

★ ★ ★

The Globalization of Technology

Gerald P. Dinneen

Whether it's a McDonald's in Moscow or a Michael Jordan poster in Tokyo, we hear a lot about the "globalization" of our political system, culture and economy. But the gap between the United States and the rest of the world also has narrowed in a way that is widely misunderstood — and unnecessarily feared.

This narrowing is in the field of technology. The technologically unipolar world dominated by the United States has given way to one in which we are, at best, first among equals. It's no secret that many products are manufactured abroad, but what's far less obvious is how much of the technology used to *make* these products is now developed overseas.

Many U.S. firms are turning to foreign engineers to design automobile parts or electrical devices. Software companies are springing up in India, Taiwan and Brazil. Researchers in Japan no longer are just copying American technology but are blazing the trail in such critical fields as microelectronics, optics and materials. The Pentagon is shopping abroad for the components of some high-tech weapon systems.

Overall, the days of American technological pre-eminence are over. This change is of profound significance when one considers how essential technology is to the economy, the military, the environment, health care and our children's future.

The trend has its good side. When Honda produces automobile engines that offer better gas mileage, or Toshiba devises clearer screens for laptop computers, American consumers benefit. But the rapid growth of foreign technical competence and competition challenges our own ability to grow or capture the high-tech industries we need to prosper.

Some observers say the solution to our diminishing technological edge is to get tougher about protecting the expertise we still retain, making it harder for others to apply our know-how. But putting a fence around knowledge in an era of fax machines and jet planes is an exercise in futility. It also misses the point that the United States increasingly has as much to gain from technology exchanges as it stands to lose. In a recent report, the National Academy of Engineering concluded that technological convergence and industrial interdependence among nations are essentially irreversible and, on balance, actually positive for the United States.

Globalization enables us to take advantage of technological resources and skills from abroad. Our automobile, steel and computer manufacturers, for instance, all have adopted technologies and management techniques from Japan. Increasing technological sophistication abroad also has opened up new markets for our own goods and services, and provided American consumers with a wider range of high-quality products at reasonable prices.

So the benefits are substantial, but only if U.S. companies and federal policy-makers stop being so preoccupied with home-grown U.S. research and development. We need to take advantage of technological advances no matter where they originate. In a high-tech, interconnected world, prosperity follows not only from creating technology but from harnessing it effectively, regardless of its national origin.

In practical terms, the United States must develop the human, financial, physical, regulatory and institutional support systems to increase its attractiveness as a location for firms to perform the full spectrum of advanced technical activities. Just as California's Silicon Valley competes with Boston's Route 128 for high-tech business, our technological community as a whole now is competing with the rest of the world.

Private companies must lead the charge, but they cannot win this battle alone. Government at all levels must play a bigger role. State and federal policy-makers should do much more to help promote technological "best practices" and commercially significant generic technologies throughout private industry. The federal government must become more aggressive about reaching an international consensus on trade, antitrust regulations and other sensitive questions that affect the flow of technology.

Over the long term, we will remain competitive in the technological arena only by investing in the public education, job training and other activities that will allow us to harness the world's expanding technology base as well as, if not more effectively than, our trading partners. Unless Americans gain the necessary skills, the rest of the world will do more than catch up to us in technology; it will leave us behind.

November 17, 1991

***Gerald P. Dinneen**, a former vice president of technology at Honeywell Corp. and assistant secretary of defense, is foreign secretary of the National Academy of Engineering.*

* * *

How to Keep Factory Jobs from Moving Overseas

Laurence C. Seifert

1. The check is in the mail.
2. Let's do lunch sometime.
3. Gee, that gift was *exactly* what I wanted.

To these dubitable statements add: "American factories are fleeing overseas to take advantage of cheap wages and benefits."

It's an argument heard all the time, especially amid the debate over the North American Free Trade Agreement. Television commercials depict Americans losing jobs to poor workers in the developing world. Economists say the United States is having trouble competing in manufacturing because its workers are too well-off and inflexible.

By and large, the argument is wrong. The problems facing U.S. manufacturing are not due primarily to low wage rates in third world factories.

A committee of the National Research Council, which I chaired, reported recently that there are more significant factors involved when companies decide to open facilities abroad. Labor costs are critical when manufacturing products are essentially commodities. But increasingly, U.S. manufacturers are looking abroad primarily to gain access to new markets, manufacturing processes, technologies and components.

Our study focused on three industries often cited as examples of the conventional wisdom: automobiles, consumer electronics and semiconductors. In each case we found other factors to be more important than labor rates in corporate decisions about where to build new facilities.

In the automobile industry, production remains concentrated overwhelmingly in companies' home markets. Ford and General Motors have a strong presence in Europe, and the Big Three also manufacture in many developing countries to meet "local content" requirements. But most U.S. auto production is done at home. By contrast, Japanese auto makers have opened major plants in North America and Europe, mainly to respond more quickly to market changes and customer requests. Toyota is producing station wagons in Kentucky for sale in Japan.

In the 1970s and 1980s, many U.S. companies did move facilities overseas to take advantage of low wages. But these were mainly simple assembly operations. Today's high-tech, automated manufacturing depends much less on human labor. AT&T has been producing consumer telephones in Singapore since 1984, but a recent study showed the operation's success is due more to access to lower materials costs than to cheaper wages. Toshiba began producing color picture tubes in New York in 1985 mainly to gain closer access to the U.S. market, and structured it to be its least costly operation.

Wage rates remain critical for labor-intensive semiconductor packaging. But even in the semiconductor industry, success depends more on manufacturing prowess, proximity to customers and political factors such as trade barriers than on low wages. U.S. firms also open foreign facilities to learn the innovative production techniques pioneered in Japan and elsewhere.

So arguing that the United States has become too expensive for manufacturing is inaccurate. Higher wages are a deterrent, but more often our country offers features that companies most desire, including a large market of affluent consumers, skilled workers, a strong technological base and a tradition of innovation.

Blaming our problems on poor foreign workers only diverts us from more germane concerns. Although U.S. manufacturers have made progress during the past few years, many still have not adapted to a world in which quality, flexibility and speed are essential. Their failure puts American jobs at risk.

The federal government, meanwhile, could be doing more to make the United States an attractive choice for manufacturers. This country needs favorable tax and trade incentives, and should avoid any restrictions that inhibit U.S. producers' access to new technology, whatever the source. It also must ensure that its workers are better educated and have the job skills that manufacturers require. Protecting the jobs of underskilled workers through political means is an equation for long-term economic disaster.

American companies sometimes have good reasons to open facilities overseas. To protect American jobs, the first step is to stop hiding behind the myth that cheap foreign wages are both paramount and insurmountable. Neither argument is true. With the right incentives, skills and resources, and with enlightened management, manufacturers based in the United States can compete with anyone. Our manufacturing problems begin at home. If we get busy solving them, we'll do just fine.

October 4, 1992

Laurence C. Seifert *is vice president of global manufacturing and engineering at AT&T.*

★ ★ ★

Designing for Prosperity

Charles W. Hoover Jr.

One reason why the economy lingers in recession, with millions of Americans unemployed, has little to do with the usual causes we hear on the news. It's that so many companies haven't figured out how to design products that people will buy.

Not "design" in the sense of the Reebok Pump sneaker having a pump shaped like a basketball, or AT&T's new laptop computer sporting an aerodynamic design. From the dashboard of the Plymouth Voyager to the soft lines of Black & Decker's Dustbuster, aesthetic concerns such as these get plenty of attention from manufacturers and consumers alike. American companies do not always succeed in designing attractive products, but they try.

Where they lag badly, however, is in a kind of design much less obvious but even more important. And their failure in this arena hurts anyone who is looking for a job, owns stocks, has a pension or cares about our country's economy.

This is "engineering design" — the process of turning a concept into a finished product, of figuring out how to make the new toaster, dress or computer as efficiently as possible. I co-chaired a committee of the National Research Council that recently examined engineering design in the United States, and we found the overall quality to be poor. Unless it improves, our companies and workers face real trouble.

Consumers see only the retail price of a product. But long before the product reaches the store, 70 percent or more of its cost is determined by its design. One reason IBM had such success with its "Proprinter" a few years ago, for instance, was that a designer named Charley Rogers led a team that found a way to build the printer from a few parts that could be assembled with simple vertical motions. The printer was not only easier to build but less likely to break. It sold like hot cakes.

Simplifying the design is not the only way to improve the

Drawing by Ned Levine

quality and speed of manufacturing. Computer-aided design and engineering has evolved from a promising concept into a powerful tool for conceptualizing possibilities and evaluating costs. New "benchmarking" techniques enable designers to help win business by setting performance standards exceeding those of competing goods. Improved accounting approaches quantify the true contribution of designers, helping them get needed resources.

More important than any specific technique, however, is simply recognizing the importance of changes like these. Our automobile and consumer electronics industries learned the hard way, by losing business to foreign competitors. Yet there continues to be widespread denial among other U.S. manufacturers that their design practices are outmoded. Because of managerial indifference rather than any technical barriers, many companies are unwilling even to try new design practices.

In an era of Toyota cars, Armani suits, Canon copiers and countless other foreign products, such complacency virtually guarantees further decline of U.S. competitiveness. What's needed is a complete rejuvenation of engineering design. Hewlett-Packard, Ford, Xerox and a handful of other U.S. companies have shown how to achieve this, implementing new design practices that yield shorter development times, lower costs and more desirable products. These pioneers typically needed between five and eight years to change their habits, and many of them now are willing to share their know-how to help other U.S. firms. But these others first need to wake up to the problem.

Companies need help. University engineering departments should be working closely with them to develop new design methodologies and to train the next generation of design engineers. Yet many engineering schools now give these concerns short shrift. Professors generally have little significant industrial design experience, understanding of current manufacturing, or contact with companies. Relevant textbooks are scarce. Research in the field is inadequate. Mechanisms for transferring new ideas from campus to companies are lacking.

For the sake of millions of American workers, all this must change. The need to revive engineering design must be recognized both in the corporate suite and on campus. If we don't learn to make better products, we can look forward to foreign companies' taking away ever more jobs and sales from Americans. Our continuing failure to take engineering design seriously is itself the perfect design for economic decay.

June 23, 1991

Charles W. Hoover Jr. *is professor of manufacturing engineering at Polytechnic University in Brooklyn, N.Y.*

* * *

Short-Term Thinking in a Long-Term World

Donald N. Frey

Recent reports indicate that spending on research and development by U.S. companies is falling while foreign firms increase their R&D efforts. The reports are an alarm bell for our future. If the current recession has taught us anything, it's that we cannot focus just on immediate profit to the exclusion of longer-term success without risking our jobs and prosperity.

U.S. businesses obviously need to be responsive to short-term challenges and opportunities. But they cannot survive over the long haul unless they also pursue new technologies and innovation. The heart pacemaker took 32 years from original concept to realization. The video tape recorder took six years — much faster but still beyond the current planning horizons of many of our companies. It's no wonder foreign competitors often cash in on these advances.

Some U.S. firms, such as those in the pharmaceutical, aerospace, chemical and food industries, have become adept at deliberately pursuing and profiting from long-term innovation. But a committee that I chaired for the National Academy of Engineering reported recently that a shorter-term focus too often remains the norm and a major source of competitive disadvantage. The recent R&D figures from the National Science Board are subject to differing interpretations, but other data indicate that U.S. firms spend relatively less than their foreign counterparts on new equipment, machinery, and non-defense R&D.

Corporate executives, who report to stockholders, can be encouraged to take a longer view. They now often are *rewarded* for acting short-sightedly.

Imagine that you are a business leader deciding whether to start a new product development that could help your company beat its foreign competitors. Initiating the project will reduce profits this year and thus cut the size of your own performance bonus.

It would take a person of considerable vision — or, even better, wealth — to accept such a personal loss in favor of the company's long-term vitality. Yet that is precisely what many executives are expected to do. They are given little incentive to commission research or other initiatives likely to blossom after they retire. Their pay and reputation are based upon what they accomplish now.

For the sake of millions of American stockholders (and workers), corporate boards should develop compensation plans that induce their executives to adopt longer time horizons. How? Managers might be given stock options that cannot be exercised for several years, incentives tied to long-term company performance, and penalties for cashing out sooner.

More-sensible compensation packages are only part of the solution. Firms also should pursue "patient" investment capital to give themselves more flexibility. They should take advantage of joint ventures that spread risk among several players, making it easier for all to pursue untested innovations.

On a national level, we need to lower the cost of investment capital. Potential measures include cutting the federal deficit, reducing capital gains taxes and "double taxation" of corporate profits, and promoting personal savings. The federal government also should improve the efficiency of its regulatory, patent and licensing procedures.

Although the recent R&D figures are sobering, there are signs that at least some companies are getting the message. One corporate board on which I sit recently had a meeting at which someone suggested that the company's R&D budget should be reduced. A fellow board member replied indignantly, "What do you mean cut the R&D budget? If you try to increase earnings by doing that you're just liquidating the company!" That's something I'd never heard before in a board meeting.

Nor until recently had I heard a board ask to review the firm's research budget, or an employee take time to describe some new machinery in the factories. In the past, the discussion rarely got beyond financial matters.

So there's reason to hope that some companies finally are focusing on these nuts-and-bolts concerns involving technology, quality and the shape of the future. But regardless of what the federal government does, our country's prosperity depends on many more firms extending their investment horizons still further. We've learned the hard way that a relentless focus on short-term profit is a shortcut to failure in a long-term world.

March 15, 1992

***Donald N. Frey**, professor of industrial engineering and management sciences at Northwestern University and former chief executive officer at Bell and Howell Co., chaired a National Academy of Engineering committee that studied "time horizons and technology investments" in U.S. industry.*

★ ★ ★

A New Partnership in American Science and Technology

Richard F. Celeste

After a spring marked by urban violence, plant closings and recession, state officials across the country are under intense pressure to come up with new ideas for revitalizing their ailing economies. One approach, which became popular during the 1980s in many state houses, is to create high-paying, high-tech jobs by nurturing local expertise in science and technology.

The same idea is now gaining favor within the federal government, which has begun looking beyond its traditional support of basic scientific research to helping companies actually make use of innovations.

The United States leads the world in scientific research. But Japan and other nations, often with the active support of their governments, have proven more adept at putting new ideas to work on the factory floor and in products ranging from automobiles to VCRs.

Although the initiatives are welcome, federal and state officials could easily trip over one another as they seek to get science and technology out of the laboratory and into the marketplace. As Adm. Hyman Rickover put it, "Trying to make things work in government is sometimes like trying to sew a button on a custard pie." This must not occur; the stakes are too high and resources too scarce.

A decade ago, state spending on science and technology was scant, going mostly for agriculture. But these investments have skyrocketed to nearly $1 billion in 1990 alone.

Some states, like Massachusetts, are providing seed money for local businesses. Others, such as Utah and Pennsylvania, have established research and technology projects that capitalize on the strengths of local universities. New Jersey is backing research based on its leadership in pharmaceuticals and electronics. Alaska is supporting projects involving everything from DNA tracking of salmon to generating energy with waterwheels. Other states are placing their bets on research parks that provide growing firms with laboratory space and other facilities.

The federal government necessarily takes a broader view, focusing less on the specific needs of any region than on the scientific and technical strength of the country as a whole. The National Science Foundation, the National Institutes of Health and other federal agencies support most of the basic research in the United States. Usually the goal is knowledge for its own sake. However, Washington also supports applied research for the military, health care and other vital fields. In addition, it has begun joining with private firms and others in supporting research aimed at commercializing innovations.

In other words, both state and federal entities are moving beyond their traditional roles. And their efforts have the potential to create new industries, jobs and hope for American workers.

Wasteful overlap must be avoided. In general, the federal government should maintain its commitment to basic science and engineering research and to meeting national needs. The states should concentrate on applied research and the diffusion of technology to meet local economic development objectives.

Rather than fighting over turf as these distinctions increasingly fade, federal and state officials should search for ways to build on each other's efforts, inventing new forms of cooperation.

Michigan, for instance, has begun taking advantage of a new center established by the state and the National Science Foundation. One new local company is using technology from the center to manufacture devices that apply diamond-thin coatings to cutting tools.

Illinois is helping local firms commercialize technology developed at Argonne National Laboratory, located outside Chicago. The University of Chicago also is involved in the effort.

The Cleveland Advanced Manufacturing Program, with support from the U.S. Department of Commerce, is helping small- and medium-sized firms to retrain employees to use advanced new equipment.

There is no simple recipe for this kind of cooperation. The main thing is for everyone — government officials, business executives, university researchers and others — to push in the same direction.

Science and technology alone cannot put an end to our country's serious economic and social problems. But they are an essential part of the solution. As the federal government and the states both seek to help U.S. companies become more innovative and competitive, they must learn to dance together instead of ignoring one another or, even worse, stepping on each other's toes.

American workers won't be helped by bureaucratic turf wars. They need jobs.

June 21, 1992

***Richard F. Celeste**, former governor of Ohio, chairs the Government-University-Industry Research Roundtable of the National Academy of Sciences, National Academy of Engineering and Institute of Medicine.*

* * *

Work and Family

Lotte Bailyn

Ever taken a sick day off work? Try this quiz. What percentage of workers in the United States do you think are offered paid sick leave by their companies: (a) 46 percent, (b) 67 percent, or (c) 88 percent?

Now a question for working mothers. Among married women with children under the age of 6, how big an increase would you guess there has been since 1960 in the number of those who have joined the formal labor force? Has the percentage: (a) gone up by half, (b) doubled, or (c) tripled?

A final question: Who generally gets more holidays and vacation days: workers in western Europe or those in the United States?

The answers suggest that the health care crisis and the continuing economic recession are not the only middle class issues likely to loom large in the 1992 presidential election. Millions of Americans also are straining with another problem: juggling work demands with family responsibilities.

The first question has two correct answers, neither of which is reassuring. Only 46 percent of employees in small firms have paid sick leave; 67 percent do in large firms. The answer to the second question is (c). Among married women with young children, the percentage in the labor force has risen from 19 percent in 1960 to 57 percent in 1988. As for international differences, workers in western Europe not only get more holidays and vacation days but also are more likely to receive extensive family leave and publicly subsidized child care.

An expert panel of the National Research Council on which I served reported recently that employee benefits have failed to keep pace with a labor force increasingly composed of people balancing work demands with family responsibilities. About half of American workers care for children, elderly parents or other family members. Fewer than a third of employees have a spouse at home full time.

Our panel analyzed data from a wide variety of sources and looked beyond political rhetoric to determine what the situation really is with parental leave, flexible hours, health insurance and other benefits. We found a clear need for more attention to the family concerns of American workers. Although more and more firms offer family-related benefits, many legitimate needs are not being met. For instance:

- The majority of employed women have no paid leave for pregnancy or childbirth, and many lack unpaid leave as well.
- Roughly 12 million children lack health insurance, even though they live in families that have at least one employed person.
- Only 15 states have enacted parental leave laws, and these laws generally exempt small businesses.

Minorities and women raising families alone are especially likely to lack benefits since they tend to work for small firms and in seasonal or part-time jobs. Smaller firms create many new jobs and are somewhat more likely to offer part-time work and flexible schedules. But they also usually pay less and have fewer benefits; more than half do not provide paid sick leave.

Many readers undoubtedly know from their own experience what these numbers mean in human terms. Millions of Americans struggle every day to perform well on the job while finding time to care for a sick child, help an elderly parent or manage family crises. Particularly for single-parent families, there is evidence that economic and psychological stress has negative development effects on children. Employers suffer, too; data suggest that family responsibilities exacerbate absenteeism, tardiness and other work-related problems.

It is clear that the current system of employee benefits in the United States is inadequate for the diverse labor force of the 1990s and beyond. Improving the situation will not be easy, especially with many companies simply seeking to survive the current recession. But employers, unions and policy-makers must devise new ways to provide workers with family leave and paid sick leave, with more flexible schedules and work locations, with greater resources to help care for children and elderly or disabled relatives, and with adequate health care coverage.

Perhaps specific solutions will emerge in the upcoming election campaign. The numbers leave no doubt, however, that millions of Americans are feeling the strain between the conflicting demands of work and family.

December 22, 1991

Lotte Bailyn *is professor of organization psychology and management at the Sloan School of Management at the Massachusetts Institute of Technology.*

8

INTERNATIONAL AFFAIRS

Beyond the Brazil Summit: Conserving Biodiversity

Peter H. Raven

During every second that President Bush and other leaders spend at this week's Earth Summit in Brazil, the world will lose nearly one acre of forest. Each day hundreds of species will vanish. When the conference ends and everyone returns home to deal with other problems, this destruction will continue.

In the 20 years since the world last paused, in Stockholm, to ponder its environmental plight, an area of tropical forest has been cleared equal to all of the United States east of the Mississippi River. A fifth of the topsoil from the world's arable lands is gone.

We must not allow these trends to continue after the politicians depart from this week's conference. Decisive action is needed to deal with global warming and other environmental problems. As a biologist I worry especially about continued destruction of tropical vegetation and species. During the past 30 years about one-third of the tropical rain forest has disappeared. An equal amount may vanish over the next 30 to 50 years, taking with it a quarter of the world's species diversity. Arid and semi-arid regions also are losing species at alarming rates.

The loss of this biological heritage is incalculable. We depend on animal, plant, fungal, and microbial species for food, fuel, fiber, drugs and raw materials. Our agricultural bounty is based on formerly wild plant and animal germplasm. Living organisms mitigate pollution.

The public demand for action to preserve forests and "biodiversity" has grown tremendously. That's good, because scientists alone cannot solve a problem that involves politics and economics as much as science. But even as we act, we must plug some significant gaps in our understanding of the problem.

A National Research Council committee, which I chaired, concluded recently that a lack of scientific information is hampering our ability to comprehend the magnitude of the loss of biodiversity, prevent further losses, and formulate sustainable alternatives to depleting resources. Answers are unavailable for seemingly simple but important questions: How many species are there? Where do they occur? What is their ecological role? What is their status — common, rare, endangered, extinct?

To date, about 1.4 million kinds of organisms have been assigned scientific names, but coverage is relatively complete only for a few taxonomic groups such as plants, vertebrates and butterflies. Most groups and many major habitats such as coral reefs or forest canopies remain poorly studied. Estimates of the Earth's total species diversity range enormously, from 10 million to 100 million. That's like calculating a fortune without knowing whether you have dimes or dollars.

Although the situation is too critical to wait for research to reveal in full detail how we may sustain biodiversity permanently, the U.S. Agency for International Development and similar bodies should move aggressively to resolve these many scientific questions.

The most basic research requirement is to gain a better sense of "what's out there." We need to know more about how biological diversity is distributed, how it is faring, how to protect it and use it in a sustainable manner, and how to restore it.

With biological resources disappearing so quickly, it also is essential to identify the most effective methods for preserving and restoring ecosystems. Genetic engineering and other scientific advances have the potential to boost agricultural production, thereby reducing the need to clear undisturbed forests. Botanical gardens, seed banks and other facilities could help preserve information on biodiversity. Yet many nations with the highest concentrations of biological diversity are crippled by persistent poverty and high rates of population growth. We must learn how to help their local institutions make things better.

The best way to deal with the problem is through a series of what might be called national biological resource commissions, which bring together the private and public sectors of countries in studying how to economically exploit biodiversity for national benefit while conserving it for future generations. The Instituto Nacional de Biodiversidad in Costa Rica is an excellent example. Similar organizations are being formed in Taiwan and Mexico. They deserve our support.

The loss of biodiversity is irreversible; species that are lost are lost forever. The potential impact of that loss on the human condition, on the fabric of the Earth's living systems and on evolution itself is immense. A delay of even five years will be too late to prevent staggering losses. When the Brazil conference ends, the real challenge begins.

May 31, 1992

Peter H. Raven *is director of the Missouri Botanical Garden in St. Louis.*

* * *

Faltering Science in the Rain Forest

Thurman L. Grove

It's been weeks since the world packed up its cameras and left the Earth Summit in Brazil for other issues. But deep in the Amazon jungle, where few politicians or journalists tread, the battle to preserve the world's biggest rain forest continues.

At the forefront of that battle are scientists racing to categorize and study the forest's plants, animals and other "biodiversity" before it disappears. These Brazilian experts have foregone comfortable careers in the cities to carry out research for little pay or recognition. They are true heroes on the environmental front lines.

With support from the U.S. Agency for International Development, I recently led a National Research Council team that visited many of these scientists. We found their situation intolerable. Their pay is pitiful, equipment is falling apart, and working conditions are needlessly dangerous. If Brazil and the international community are serious about saving the rain forest, they need to help these scientists immediately. Otherwise, the world may lose its last chance to learn about — and perhaps preserve — this precious ecosystem, which helps cleanse the world's atmosphere.

Two Brazilian institutions are primarily responsible for our knowledge of the Amazon forest and its inhabitants. The Emilio Goeldi Museum in Belem, near the mouth of the Amazon River, was established in 1866. The National Institute for Research in the Amazon, in the middle of the forest, was established in 1952. Together they house irreplaceable collections of biological and cultural materials, much as the Smithsonian Institution serves as the treasure trove for our country.

Both Brazilian institutions are in jeopardy. The Goeldi Museum attracts a half-million visitors annually to see the excellent zoo and educational programs. But money now is so tight that the veterinarian who cares for the animals has no running water in his clinic. Caretakers prepare food for the animals in a condemned building with no refrigeration — this in a tropical country where food spoils quickly. There even are shortages of the stainless steel pins used to mount butterflies.

At the forest research institute, meanwhile, the walls are scorched because the electrical circuits cannot handle modern equipment. There is no safety equipment to protect against noxious fumes and other dangers. The library lacks money to buy books, much less to connect with "on line" scientific data bases via computer.

Most worrisome is that priceless collections of Amazonian plants at both institutions are imperiled by a lack of reliable air conditioners and dehumidifiers. From a biological standpoint, this is like taking the Declaration of Independence out of its protective case at the National Archives and placing it outside in the afternoon sun. The Amazon is home to countless species, some of which undoubtedly could yield new medicines and other products to improve life not only in Brazil but in the United States as well.

The people we met at these institutions were very dedicated. They live thousands of miles from their professional colleagues and extended families. Many could earn more money with less trouble in Sao Paolo or Rio de Janeiro. Yet they stay, trying to do scientific work that is essential to preserving our planet's precious biodiversity.

Their plight is not just Brazil's problem but our own. If the Rio conference taught the world anything, it is that environmental problems such as saving tropical rain forests extend beyond national borders. Brazil contains more than a third of the world's tropical moist forests, but is clearing the land at an alarming rate. Unless this changes, the entire planet could feel the consequences.

The United States has pledged to contribute to an international effort to help Brazil reduce deforestation. But it has not yet contributed the money, much less taken any leadership in bolstering these faltering scientific efforts. At the least, the United States should support international efforts aimed at rebuilding the Amazonian institutions and helping the experts who work there.

While others make speeches about the Amazon, the scientists in the Amazon are facing the daily rigors of living in a hostile environment characterized by heat, humidity and disease. They are the ones doing the hard work. If we meant what we said at the Earth Summit about helping other countries preserve the Global Commons, here's a chance to prove it.

September 6, 1992

Thurman L. Grove *is professor of soil science and assistant dean of the College of Agriculture and Life Sciences at North Carolina State University.*

* * *

The Perilous State of Science in the Former Soviet Union

Frank Press

Few people know more about earthquakes and related natural phenomena than V.I. Keilis-Borok. He is one of Russia's most renowned scientists and a foreign associate of our own National Academy of Sciences. I conducted research with him for more than 20 years.

His current salary is equivalent to $15 a month.

My friend Volodya is not alone. Many senior researchers in the former Soviet Union are being paid less than Moscow's bus drivers. They lack the funds to run their computers, restock their laboratories or pay for scientific journals. University faculty are bringing their own light bulbs to lecture halls. The largest scientific publishing house in the former Soviet Union reportedly has turned to publishing a book on astrology to raise cash.

All this has attracted attention in the United States, mainly because of fear that former Soviet weapons experts may go to work for Iraq or other countries eager for their expertise. But that danger is only the most obvious reason why our country should greatly expand its efforts to assist scientists and others whose expertise is essential for the former Soviet Union seeing its way successfully to a future that so heavily affects our own.

Our Academy has been in almost daily contact with colleagues in Moscow, St. Petersburg and other scientific centers. They have cited example after desperate example of how world-class research is being suspended due to a lack of budgetary support and an exodus of experts.

It is hard to overstate the magnitude and importance of this teetering scientific enterprise. The former Soviet scientific system must remain vital if reform efforts are to succeed. Russia and the other states need technical expertise to achieve economic growth, clean up the environment, expand agricultural output, provide health care and meet other essential goals.

Since these nations are home to some outstanding research groups, it also is in our own interest to collaborate with them. Firms from Germany, Japan and other countries already have begun searching out commercial targets, with strong backing from their governments. U.S. companies and government should be seeking similar business alliances before the best opportunities disappear.

The former Soviet Union also is home to many research facilities, data sets and other scientific resources whose value is less commercial but perhaps even more profound. Years of study that could benefit the world may be damaged or lost.

At a recent meeting at our Academy requested by D. Allan Bromley, assistant to President Bush for science and technology, more than 100 U.S. scientists and engineers warned that time is of the essence. Our government must act immediately and aggressively to help reorient the basic science and technological capabilities of the former Soviet Union and stem the brain drain of its experts.

Congress and the administration deserve praise for their recent efforts to provide funds to help dismantle the nuclear weapons complex within the former Soviet Union and provide new opportunities for its weapons scientists. But we must not ignore the many talented scientists who did not participate in Soviet military research. The best of these civilian scientists should be recruited for joint projects with our own experts. Programs are in place to promote this.

Rep. George Brown, chair of the House Committee on Science, Space and Technology, has called for the establishment of a binational science foundation to pay for basic research in Russia and elsewhere. Other proposals would refurbish former Soviet labs and libraries. A number of U.S. companies, meanwhile, are prepared to consider commercial, non-military business arrangements in the former Soviet Union if given the go-ahead from the Defense Department.

These and other initiatives are a small fraction of the total funds being considered to help prevent economic collapse and other turmoil in Russia and the other republics. The total is a large sum at a time of economic recession in our own country but small when compared with the burden we may face if economic recovery and democratization fail in the former Soviet Union. Stopping the free fall of science and technology there is an investment in our own country's security, economic and other interests.

Russia and the other republics have a proud scientific and technological tradition. They have given the world the periodic table, the first orbiting satellite, and many other breakthroughs. If they now become scientific backwaters, we all lose.

May 10, 1992

Frank Press *is president of the National Academy of Sciences.*

* * *

Guatemala: Attacks on Scientists and Research

Eliot Stellar and Carol Corillon

On a radiant morning in Guatemala several months ago, tourists blithely explored the colonial city of Antigua and shopped in the colorful Indian market in Chichicastenango. Although Guatemala is just a few hundred miles south of the Texas border, most Americans are unaware that some 40 thousand people, mostly Mayan Indians, have been killed or made to "disappear" for political reasons.

We spent that bright day in Guatemala City learning the chilling facts of the murder of anthropologist Myrna Elizabeth Mack Chang, and expressing concern to government officials about her death and the deaths of more than 30 other scientific colleagues in Guatemala who, during the past 10 years, have been murdered for political reasons or abducted and never seen again. We used the Mack case, which has now entered a crucial phase, to symbolize the continuing assault in Guatemala on scientific researchers — an assault that should concern anyone who cares about scientific freedom.

We visited Guatemala as part of a five-person delegation from the Committee on Human Rights of the National Academy of Sciences and the Committee on Health and Human Rights of the Institute of Medicine.

Mack conducted field work on Indian peasants displaced during the 1980s because of the guerrilla war. She wrote about the effects of the government's repatriation policies on their lives and about their treatment at the hands of Guatemala's powerful army and civil patrols. She gave the Indians a voice. As a result, her own voice was silenced.

On the evening of September 11, 1990, Myrna Mack was approached in Guatemala City by one or more assailants. An extensive struggle ensued and she was stabbed 27 times. Her murderers fled with her field notes.

Although Guatemala's guerrilla war has subsided, political killings have not. Two weeks before our visit, Manuel

Peña, a history professor at the University of San Carlos, was shot dead by at least two assailants using automatic weapons. Peña worked with the underprivileged and internally displaced.

These assaults have a chilling effect on other researchers in Guatemala. As one scientist who worked with Mack told us: "We were not aware of the consequences of our academic work . . . and now we are too frightened to carry the work forward."

The police officer who investigated the crime was himself killed, in front of his family, shortly after implicating the military in his homicide report. A former soldier in the intelligence branch of the Presidential High Command has since been charged with Mack's murder and is under arrest. There is speculation that other members of the military also were involved in her murder.

At the request of the Mack family, the courts have called a number of witnesses, including a former president of Guatemala and military officers. The army has complied with a court request for a large number of records that might shed light on the case. After passing through a dozen or so courts, the case now is at a crucial stage known informally as "the trial phase." That the trial has progressed so far is a hopeful sign, although, as with the peace talks with the leftist guerrillas, progress has been distressingly slow.

We spoke with Guatemalan government and military officials who expressed support for the policy of President Jorge Serrano Elias to improve human rights abuses and reach a just resolution of the case. We visited Myrna Mack's parents, siblings, and 18-year old daughter, who have been unrelenting in their demands for justice despite potential repercussions to themselves. They told us that the U.S. ambassador to Guatemala has encouraged them and maintained firm pressure on the Guatemalan government to resolve the case. Mack's friends and scientific colleagues also have been tireless in keeping the case alive.

Coming as we did from a country where free scientific inquiry is taken for granted, we were deeply impressed by the difficulties and dangers that Myrna Mack faced out of dedication to her work, and by the courage of her family and friends following her murder. The rest of the world, particularly fellow scientists, should join them in demanding that those who murdered Myrna Mack, as well as those who planned the crime, be prosecuted and punished. Thirty years of murder with impunity in Guatemala must be reversed.

September 27, 1992

Eliot Stellar**, chair of the department of cell and developmental biology and University Professor of Physiological Psychology, University of Pennsylvania, Philadelphia, chairs the Committee on Human Rights of the National Academy of Sciences.* ***Carol Corillon *is the committee's director.*

* * *

The Next Refugee Crisis and the U.S. Response

Carl E. Taylor

Last year new waves of desperate faces from Iraq, Haiti and Hong Kong joined the familiar starving babies from Somalia and Sudan. It's too soon yet to know where this year's refugees will originate. The one thing that *is* certain is that new refugees will appear somewhere.

The number of refugees and displaced persons worldwide has increased almost geometrically to between 30 million and 50 million. Yet because the problem seems relentless, one can easily feel hopeless.

Even with our current preoccupation with domestic issues, we must resist such ethical numbness. The problem of refugees is changing in ways that demand new solutions. The victims increasingly are not able to cross international borders but are "internally displaced persons," mostly women and children, unable or unwilling to leave their country. Even though the threat of war between the superpowers has declined, changing world politics are exacerbating local conflicts and ethnic hostilities.

Many of these displaced persons face appalling conditions. In 1988 the death rate of one internally displaced population in Sudan was 60 times greater than that of non-displaced Sudanese. The horror of thousands of Kurdish children dying needlessly from dysentery and pneumonia on bleak mountains was similarly grim.

Helping displaced persons can be much more complicated politically and logistically than assisting international refugees. The 1967 United Nations Protocol on Refugees was designed to help those who have crossed borders. To avoid infringing on national sovereignty, however, no protocol exists to prevent mass human rights violations among internally-displaced populations in places such as Afghanistan, Angola, El Salvador, Ethiopia, Guatemala, Iran, Iraq and the former Soviet Union. During the recent refugee crisis in Liberia, for example, many relief agencies said they were unable to help because the problem was an internal one.

Another basic inadequacy with the current relief system is that it is designed to deal with sudden emergencies. But increasingly, refugees and displaced persons require long-term assistance. Some Cambodian refugees have now lived in Thailand for 15 years. Thousands of Palestinian youngsters have spent their entire lives in refugee camps. These people have needs that go beyond food and emergency care.

Mental health care, for example, might seem like a luxury for refugees. But one recent study of children in Mozambique, victimized by war and displacement, painted a bleak picture. Among 500 children averaging 12 years of age, 77 percent had witnessed death, 88 percent had witnessed torture, and 51 percent had been tortured themselves. More than half had been kidnapped from their families. Most of these children had sleep disorders, body pains and various behavior problems, symptoms of what is now called post-traumatic stress disorder. Many had become ruthless boy soldiers.

In cases like these, mental health care is essential. Other

kinds of services — such as basic education or job training — also are needed as short-term crises become long-term impasses.

Then there is the explosive question of repatriation. Many Americans felt outrage when officials in Hong Kong forced Vietnamese refugees onto airplanes bound for home. Yet now our own country is considering similar deportation of refugees from Haiti. Most refugees around the world want to return home, but effective approaches to repatriation still must be developed and evaluated.

Overall, a great deal has been learned about how to save lives and promote health among refugees and displaced persons. But this knowledge, which ranges from choosing the best sites for refugee camps to providing security and psychological support, has not been applied consistently, despite the indefatigable efforts of aid workers.

It is time to develop a more effective U.S. response. At a meeting I chaired for the Institute of Medicine of the National Academy of Sciences, experts said a new framework of international cooperation could make much better use of limited resources in helping both refugees and displaced persons. Recent decisions in the United Nations are establishing improved mechanisms and new authority to coordinate relief efforts and promote future development.

These efforts show promise, and we can be hopeful about the future. Although the numbers of refugees and displaced persons is increasing, so is our capacity to change course and provide assistance more effectively. To save lives, however, we must sustain the political and moral commitment to put this growing knowledge to work.

February 2, 1992

Carl E. Taylor *is Professor Emeritus at Johns Hopkins University.*

* * *

The Science of Middle East Peace

Zehev Tadmor

Mention "the Middle East problem" and most people think of the peace talks and other political developments. But here in the region there is an equally pressing problem that, since it has little to do with politics, gets far less attention. It is the need to expand water supplies, protect the environment, boost agricultural yields, control disease, and improve the standard of living by taking advantage of science and technology. If Israeli and Arab experts began working together more actively on these problems, they could ease many of the region's underlying pressures. Who knows? They might also inspire greater cooperation in the political arena.

Before the Gulf War, for example, the hottest issue in the Israeli newspapers was the lack of rainfall. Water levels of major lakes were falling and the country's groundwater resources were threatened.

The immediate cause of the shortage was drought and excessive use of water resources. But a more decisive application of science and technology could have eased the problem by reducing the cost of desalinating water, improving irrigation, allowing waste water to be reused more effectively and preventing the intrusion of saltwater into the water supply. Although Israel has been a leader in many of these fields, there is great potential to do more. Similar approaches could be applied in neighboring Arab nations.

If global warming causes a shift in world rainfall patterns, the need to expand water supplies in the region could become even more acute. Regardless of their nationality, local experts should be working together to avoid such a calamity. Indeed, many believe that issues related to water are more of an incentive for collaboration than for conflict.

For the past several years, scientists from Israel and Egypt have begun working together on projects in agriculture, marine science, and health through the Middle East Regional Cooperation Program. Experts from Morocco and Jordan are expected to join the effort shortly. However, the opportunity exists to do much more.

All of the nations in the Mideast, for instance, share abundant sunlight, which can be harnessed to produce electricity and reduce the use of environmentally damaging fossil fuels. Israel, a leader in constructing solar energy facilities, could share its technology with others while working with its neighbors on large regional projects and new advances.

Technical experts in the Middle East also should be working together on mathematics, space, psychology, theoretical physics, the social sciences and technology to ensure that the whole region moves forward to a better life in the 21st century.

It helps to recall that Haifa is only 78 miles from Beirut, 82 miles from Amman and 90 miles from Damascus. Those distances are but a fraction of the distance between hundreds of U.S. research centers. Of course, even short distances can become huge when blocked by fear and hatred. Scientists alone cannot end the passions that inflame the Middle East. But because science is inherently universal and uncontained by boundaries and nationalities, scientists and other technical experts can provide a model of how former enemies might work together.

More than 2,000 years ago, the great Egyptian city of Alexandria was the world's leading scientific metropolis, a place where East and West joined together in the pursuit of knowledge. It contained the greatest library of the ancient world, one that attracted scholars such as the Greek mathematician Euclid. This legacy of scientific cooperation in the Middle East needs to be revived.

My university, the Technion, opened its gates in 1924. Its first catalog stated that the university was established to educate people both here and in neighboring countries to help solve technical problems. This vision, I hope, may yet become true.

One of the most talented students I had was killed in action in the Yom Kippur War. He was not alone. Too many future scientists and engineers on *both* sides of the Middle East conflict have died before having a chance to help solve the region's many problems. Their potential to serve as a unifying force in a region now split apart by enmity and nationalism has been largely overlooked. The Middle East problem is not entirely political. It's time we gave scientists and other technical experts more of a role in solving it.

June 14, 1992

Zehev Tadmor*, a chemical engineer, is president of the Technion, Israel's leading university of physical science and technology.*

* * *

The Unwelcome Return of Malaria

Charles C.J. Carpenter

Imagine that every person in the United States — all 250 million or so — woke up tomorrow with a disease that causes high fever and chills, sometimes progressing to kidney failure, shock, coma and death. Now add another 50 million people to the total and you begin to grasp the dimensions of a worldwide health threat that requires not imagination so much as recognition.

Its name will surprise you: malaria.

Almost 300 million people in the world today are infected with malaria, and about 1.5 million of them — mostly young children — die each year from the disease. More than 2 billion people are at risk. In many tropical areas, malaria is out of control.

If you think this is a problem only for developing countries, think again. About 7 million U.S. citizens travel abroad every year to countries in which malaria is endemic. Anyone booking a trip to parts of Kenya, Thailand, and other popular destinations should be concerned about it. More and more Peace Corps volunteers are infected. Our military lost more days of work among its troops to malaria than to combat injuries during the Korean and Vietnam Wars.

Even within our borders, malaria rates are increasing, although recent cases have been treated successfully. San Diego has had several outbreaks. New Jersey had one. My own tiny state of Rhode Island had 25 reported cases last year, compared with four in 1984.

A committee of the Institute of Medicine of the National Academy of Sciences, which I chaired, reported recently that this already staggering toll is getting worse. Once thought to be nearly under control as the result of insecticide spraying and other measures, malaria has made a dramatic resurgence during the past two decades.

The disease, which has plagued mankind for centuries, is caused by a parasite that enters the blood through a mosquito bite. Malaria actually is not a single disease but has many variations. There is no vaccine of proven effectiveness. Many of the drugs used to prevent and treat the disease, notably chloroquine, have lost their effectiveness as the parasites have developed resistance.

Previous efforts to bring the disease under control have emphasized particular strategies, such as eradicating mosquito breeding areas or developing a vaccine. It now is clear, however, that this search for a magic bullet ignores the tremendous diversity of the disease. No single strategy will be applicable in all malarious regions. What's needed is a broad and diversified effort combining both malaria research and control.

On the research front, a priority of the U.S. government should be to work with other scientists around the world to develop an effective vaccine. Research efforts also must include strong support for the development of new drugs to prevent and treat the disease. Significant gaps still exist in our basic understanding of the disease. It is not known, for example, why certain people succumb to it while others do not.

At the same time, there is an urgent need to improve malaria-control programs in the field. Much illness and death could be prevented through improved education, diagnosis and treatment. People should be encouraged to use bed nets, screens and mosquito coils to protect themselves from mosquitoes. They need help with low-cost projects like draining and filling stagnant pools of water. Higher-cost efforts such as widespread spraying of insecticides also should be pursued when appropriate. A shortage of research scientists and public health experts familiar with the disease must be addressed.

The United States is the biggest single contributor to malaria research and control activities; the Walter Reed Army Institute of Research has the world's most successful antimalarial drug development program. Yet government support for U.S. efforts has lagged over the past several years, and there is a lack of coordination among the federal agencies supporting malaria research and control activities.

We have to do better. Unless things improve soon, the outlook for malaria control is bleak. The annual number of deaths is equivalent to the population of some U.S. states, and many millions of people are ill. This is not only a foreign aid issue but a threat to our own domestic public health. This ancient disease is back with a vengeance, and we need not only to recognize but to confront it.

November 10, 1991

Charles C.J. Carpenter *is professor of medicine at Brown University.*

* * *

Creating A Better Atmosphere After the Earth Summit

Robert M. White and Deanna J. Richards

Although there was discord aplenty and widespread criticism of the United States, the recently completed Earth Summit in Brazil accomplished much in setting up an international framework for action.

Now comes the hard work of achieving the goals reached in Rio. More than any other industrial nation, the United States needs to follow through and show its commitment to working with developing countries on specific environmental problems. No issue offers a greater chance for success than eliminating chlorofluorocarbons (CFCs) and the other industrial substances that are depleting the earth's ozone layer.

The United States and other industrial nations now consume as much as 90 percent of the world's ozone-depleting substances. Yet use of the substances by developing countries in refrigerators, air conditioners and other products is growing rapidly. China seeks to put a refrigerator in every home. Like other countries, it needs refrigeration not only for family kitchens but also for jobs such as keeping vaccines cold.

In 1987 most of the world's nations agreed in Montreal, with U.S. leadership, to phase out CFCs. Three years later, as scientific evidence of ozone depletion mounted, they accelerated the process and established a multilateral fund in the World Bank to help developing countries deal with the problem. The United States has contributed consistently to promoting the use of new alternative technologies.

Although it has been less than a decade since scientific evidence confirmed that CFCs and other substances are depleting the earth's ozone layer, companies in the United States and elsewhere have mounted innovative efforts to develop alternatives that cause less environmental harm. Progress has come quickly. Experts from around the world agreed at a recent National Academy of Engineering work-

Artwork from *The Fort Worth Star Telegram* by Kevin Kreneck.

shop that the remaining technical problems are less daunting than the political and economic questions of how to transfer the emerging technologies to others who need them.

Developing nations will find it difficult to pursue these alternatives on their own. The United States, by contrast, is well situated to share its growing expertise. Only by doing so can it protect the stratospheric commons. From Sri Lanka to Sao Paolo, developing countries are almost certain to increase their production and use of ozone-depleting substances in the near term, although they have agreed to phase out these substances eventually. They need technical guidance and resources to do so more quickly.

Producing the alternatives is much more complex chemically than the one-step process used to make CFCs and similar chemicals. It can cost five times more to build a factory for the alternatives and three times more for some of the ingredients. Developing countries such as India and China already have made massive commitments to CFC production and are reluctant to see these investments come to naught.

It is unrealistic to expect developing countries, many of which can barely afford to produce refrigerators and other goods with the old methods, to convert to the alternatives on their own. The international community has promised assistance, which probably will need to be supplemented. Here is where the United States can exert leadership and weigh in to good effect.

Success in eliminating ozone-depleting chemicals may rest less with technological hardware than with "softer" approaches. For example, developing countries need to gain much easier access to information if they are to learn about the latest technologies. Other problems such as language differences, a lack of training mechanisms, and inadequate local skills and infrastructures all make it hard for developing countries to take advantage of alternative technologies. Many of the countries also need help putting market forces to work and providing private industries with appropriate incentives to invest in the production, distribution and recycling facilities needed to move away from ozone-depleting substances.

The United States must play an active role in helping build these local capacities and in promoting policies that will unlock private sector investments. The fact is that no country has done more than ours to promote international cooperation in arresting global stratospheric ozone depletion. That environmental leadership was called into question amid the larger events in Rio. Now we have the opportunity to counter that impression — not with rhetoric but with action.

July 5, 1992

***Robert M. White** is president of the National Academy of Engineering. **Deanna J. Richards**, senior program officer at the Academy, helped organize an international workshop on eliminating ozone-depleting substances.*

* * *

Ravages of Nature, Disasters of Mankind

Lawrence K. Grossman

Unless our attention is riveted on the tragedy of a sudden disaster like the earthquake that just hit Southern California, it is almost impossible to get anyone to focus on all that can be done to reduce the terrible toll of nature's ravages.

Unfortunately, whenever flood, fire, hurricane, tornado, volcanic eruption or earthquake does strike, television coverage usually is so apocalyptic that we are left with the hopeless feeling that we can do nothing to reduce nature's toll. We see the worst scenes repeated over and over again, of buckled bridges, caved-in highways, collapsed houses, out-of-control floods and fires, and people in agony. And we are confirmed in our sense of fatalism about the unpredictable natural forces that periodically uproot lives and damage large parts of the countryside.

Ironically, today we can do far more to limit the losses caused by the rampages of nature than we can to control the often cruel rampages of our fellow human beings. While we cannot seem to meliorate ethnic and religious conflicts abroad or bring racial harmony to our own cities, there is no doubt that we can dramatically reduce the damage caused by floods in South Asia, hurricanes in the Pacific, earthquakes in Mexico and other natural disasters, especially those that strike our own country.

Recent studies of the 1989 Loma Prieta earthquake in the San Francisco area and Hurricane Hugo, which hit the Carolina coast the same year, have shown that science and technology now make it possible to minimize the destructive consequences of even the most hazardous events of nature. Remarkably, in view of the millions of people who were affected, a total of only 90 were killed in those two natural disasters. Given their magnitude, the toll could have been infinitely worse and, in earlier years, certainly would have been.

Nevertheless, despite great advances in hurricane warning technology, earthquake-proof building design and emergency management, Hurricane Hugo and the Loma Prieta quake still caused losses of more than $50 billion. More than a year later, thousands remained homeless.

Between 1964 and 1983, natural disasters killed nearly 2.5 million people and left 750 million injured, homeless or severely harmed, a terrible toll that could have been far less severe if proper precautions had been taken.

A report issued last year by a committee of the National Research Council offers a comprehensive program on how to reduce loss of life and property from natural disasters. Progress could be significant. As the report makes clear, "The United States is extremely vulnerable.... Direct losses from natural disasters in [this country], currently averaging $20 billion per year, continue to escalate."

I serve on that committee with a remarkable group of experts drawn from many disciplines — meteorology, engineering, seismology, volcanology, hydrology, and emergency management, among others. I am not a scientist but a broadcast executive and I was stunned to learn just how wrong my own fatalistic thinking about natural disasters turned out to be. Like most people in the world, I assumed little could be done to mitigate the consequences of nature's wrath.

As the report points out, to save life and property we need a multi-pronged effort to identify the hazards that each of us faces in our own region, then make our families aware of how to protect against them. We need early prediction and warning systems, and organized community preparedness for emergency response. We need to pay attention to nature's ability to overrun certain vulnerable areas and to wreak havoc with fire, flood, wind, mud, lava and landslides. We need to design our lifestyles, buildings, roads and bridges to respect nature's power. And we need to know *before* disaster happens, not after, how to respond and then cope with recovery and reconstruction.

The lesson I learned from my own service on this committee is one everyone else should learn. Despite all the horrors you see on television when natural disaster strikes, it need not inevitably be calamitous. Individuals can act to protect their lives and homes, and those of their families. Communities and nations can do the same.

By dint of remarkable research and long experience, technologies and practices now exist that can reduce disasters' toll dramatically. It would be a shame to wait for still another Big One to hit before we finally get around to taking full advantage of them.

July 12, 1992

Lawrence K. Grossman *is a former president of PBS and NBC News.*

9

LOOKING TO THE FUTURE

Individuality and the Brain

Gerald M. Edelman

We may soon see a headline, "Machine Beats World Champion Chess Player." Indeed, it is likely that we will be able to design a computer to carry off this feat. Does such a machine have a mind of its own? Is our brain just a machine in a machine-like world?

Some scientists have suggested, in fact, that the brain is like a computer. Others have concluded that computers can think.

If all this makes some humans uneasy, there is comforting news. Brain science and modern physics both point to the conclusion that neither the universe nor the human brain are deterministic machines. Although the computer is the most significant invention of the 20th century, it is not a thinking object comparable to our own brains.

Instead, computers are powerful logic engines that operate under precise instructions. They do not have bodies, are not conscious, and cannot function without an explicit program.

In contrast, brains are *not* primarily logic engines. They are structures that have evolved to deal with novel events. Given how many things your brain must deal with while you drive home from work, it is not surprising that the brain is the most complicated object in the universe. If you counted at a rate of one per second all of the connections in the part of your brain responsible for consciousness, you would finish counting 31 *million* years from now.

The finest wiring of the brain is individual; no two brains are alike. Brains allow their animal owners to categorize and act on events that cannot be foreseen. Computers with this much individual variation could not even run their programs. A brain memory, unlike the memory in a computer, is creative, variable and dependent on context.

Computers, being non-conscious, cannot develop a language with changeable meanings that refer to things or events. Programmers assign the meanings before and after they run their programs. Meanings developed in and by brains, by contrast, have poetic capabilities. They depend on the bodies of which they are part and on the circumstances these bodies encounter. It is unlikely that a computer-waitress would say to another computer-waitress, "The ham sandwich just left without paying."

A rich brain works like a cross between a jungle and a map. The variations in a brain's structure and function help determine which neural circuits are most fit. Just as individual animals are favored in Darwin's theory of natural selection, so individual groups of neural cells in a brain are enhanced in their connectivity if their activity results in rewarding behavior. Maps between these selected cell groups lead to rich responses, transforming past signals in novel ways. Solid evidence supporting these ideas is now beginning to accumulate.

These observations have deep implications for human concerns, particularly if each brain is necessarily individual and if each of its responses is part of a historical reaction to novelty. If these notions turn out to be true, then every act of perception is in part an act of creation. Every act of memory is in part an act of imagination. These are not acts to which computers can aspire.

Although neuroscientists like myself depend heavily on the power of computers to help us model the brain, the computers themselves are poor models for what is going on in our heads and bodies. Used well, they can help us find out about human nature and its bases. That is a sufficiently valuable role without exaggerating their capabilities. Together with the avalanche of discoveries in neuroscience, the proper use of computers will provide an exciting view of what it truly means to be human.

The Enlightenment of the 17th and 18th centuries had a hard time reconciling human freedom with a machine-like model of the world. Now we are recognizing that, far from operating like machines, individuals have identities that are profoundly unique. By revealing this, brain science is laying the ground for a new Enlightenment. President Bush has declared the 1990s to be the Decade of the Brain. Properly supported, it may be expected to yield discoveries that justify a century of freedom.

May 3, 1992

***Gerald M. Edelman**, a Nobel laureate, is director of the Neurosciences Institute and chair of the department of neurobiology at Scripps Research Institute in La Jolla, Calif. This article is adapted from his book* Bright Air, Brilliant Fire *(Basic Books).*

* * *

Mapping the Human Brain

Joseph B. Martin

The fastest supercomputer is no match for the human brain. Simply reading this article will involve billions of the reader's brain cells, communicating with each other across vast, intricate, yet precisely coordinated connections. Despite great strides in research, the human brain remains mysterious.

Our failure to comprehend many brain functions is taking a staggering toll. Millions of Americans suffer from debilitating diseases such as epilepsy, multiple sclerosis, strokes, and Alzheimer's and Parkinson's diseases. More than 60 million suffer from depression and other mental illnesses; another 20 million abuse alcohol or drugs. Still others battle the effects of head and spinal cord injuries.

These and other problems occur, at least in part, because something goes wrong inside the brain. Perhaps an imbalance of the brain's chemical transmitters leads to depression. Cells may misfire in the brain's cortex, causing epilepsy. Or a small group of cells in the brain may die, producing Parkinson's disease.

When compared with what we know about treating infections or heart disease, our understanding of the brain remains meager. As a committee that I chaired for the Institute of Medicine concluded in a recent report, this ignorance about brain function retards advances toward effective treatments for many medical problems.

President Bush and Congress have declared the 1990s to be the Decade of the Brain, making this a propitious time to take stock of what we know — and what we need to know in the future. To focus this effort, a new initiative is needed to map the brain.

Limited maps of the brain already exist. But new imaging devices, computer graphics, and other technological advances make it possible to develop three-dimensional, computerized maps of extraordinary sophistication. For researchers, this would be like trading in a one-sheet map of a city for an atlas showing the internal dimensions of individual houses. A researcher interested in an area of the brain involved in memory, for example, could obtain a computerized cross-section of the area and zoom in to view local connections and synapses.

The utility of such computerized displays of information can be illustrated by four areas of brain research. The first is vision. Blindness and disturbances of vision often are caused by stroke and other brain injuries. However, since vision involves more than 300 pathways within the brain, these problems now are difficult to understand. Intricate computer simulations, anatomical and chemical maps, and other tools might help unravel them.

Substance abuse is another critical problem. Researchers now know, for example, that cocaine exerts its activity at specific brain receptors that normally mediate cell-to-cell communications. If synthetic substances could be designed to satisfy an addict's craving for drugs by safely binding to these receptors, millions of addicts might be helped.

A third area needing more research is pain. Normal pain serves as a useful warning of injury, but many people have pain that is pathological or abnormal. Since pain involves virtually every region of the brain, as well as the spinal cord and peripheral nerves, a better map could help researchers figure out why it causes so much misery.

The last example is schizophrenia, which now afflicts 2 million Americans. Scientists studying this tragic mental illness are focusing on biochemical and other abnormalities in the brain, as well as on potential genetic or environmental causes. Whether their interest is the distribution of the brain chemical dopamine or the location of specific genes, a detailed map would be helpful.

A brain mapping initiative that helped incorporate important computer technologies into brain research could begin with an annual budget of $10 million — less than 1 percent of the U.S. neuroscience research budget. An advisory panel of neuroscientists and computer experts could guide the effort and ensure that findings are shared widely via computer networks.

A century ago, Earth mappers never dreamed of the satellite photos available today. So it is now with our new ways to map the brain, and the need to proceed is compelling. More people are hospitalized in the United States with neurological and mental disorders than with any other major disease group, including cancer. We must chart the vast biological frontier that is the human brain.

July 7, 1991

Joseph B. Martin *is dean of the School of Medicine at the University of California, San Francisco.*

⋆ ⋆ ⋆

Gene Therapy: No Longer Just a Concept

Richard B. Johnston

This is the fourth time that my patient, two-year-old Christopher, has been hospitalized with a life-threatening infection. His parents, Jan and Steven, have watched helplessly as their son suffered from recurrent pneumonia, massively enlarged lymph glands, and now an abscess around his liver. I feel frustrated because, as his pediatrician, I can treat Christopher's acute infection and reduce the likelihood of further infections, but I know there is no cure for his disease.

Christopher has a rare, inherited immune deficiency called chronic granulomatous disease (CGD). In this disorder the white blood cells that normally eat and kill invading bacteria and fungi are abnormal. Affected children develop severe infections and sometimes die.

When I gave Christopher's diagnosis to his parents, their confusion and vulnerability were painful to see. What does it mean to have a gene defect like this? Would they gain or lose by knowing whether one or both of them carried the defective gene? How would this knowledge help their son? What about their future children? Unfortunately, as a specialist in immune disorders, I have faced these questions many times from couples like Jan and Steven.

Recently, however, the answers have changed. The defective genes that cause CGD and many other inherited birth defects have been identified. Knowing the specific gene defect allows us to predict possible future problems in the affected child and what the odds are of having another baby with the same disorder.

This new knowledge can be applied not only in diagnoses but also in treatment. In 1990, a four-year-old girl with another rare, fatal, inherited immune disease called adenosine deaminase deficiency made history as the first person to receive an experimental treatment known as gene therapy.

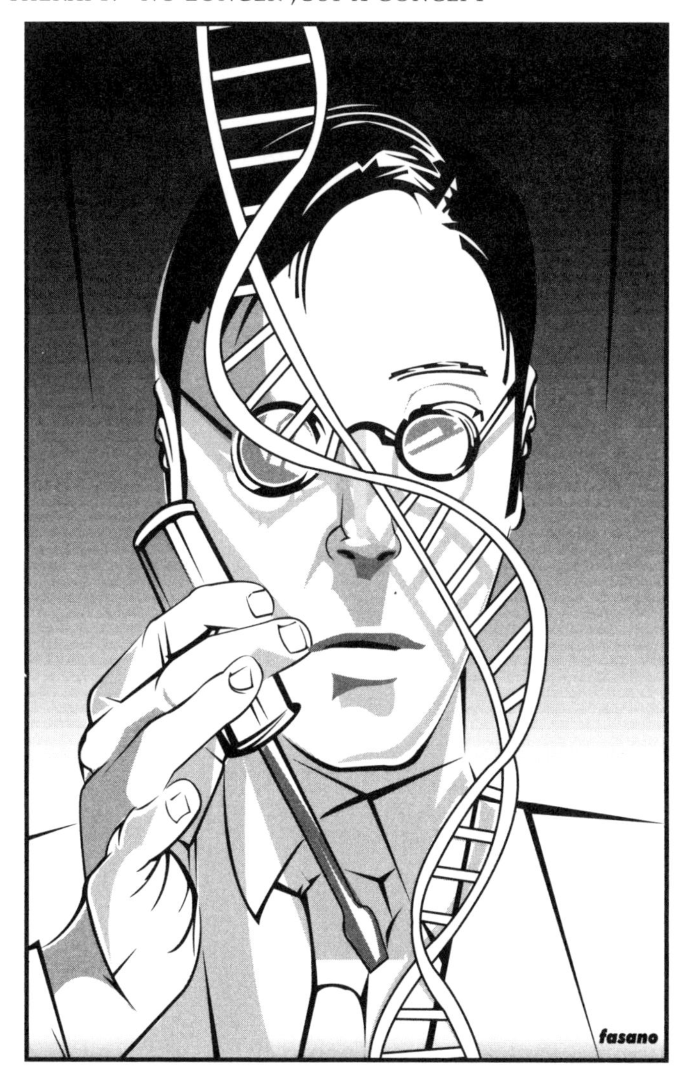

Drawing by Dana A. Fasano
The Orlando Sentinel, Fla.

Since then, a second girl with the same disorder has been treated, and the condition of both girls is greatly improved.

Gene therapy involves inserting copies of a normal gene into specific cells from the patient's body. Various tricks are used to accomplish this — for example, using "tamed" viruses to penetrate cells and implant the healthy gene. The patient can then receive a transfusion of his or her own cells, complete with healthy copies of the missing or faulty gene.

As the government's Human Genome Project proceeds to identify all human genes, it will become possible to develop gene therapy for a growing number of the approximately 3,000 known genetic disorders.

However, opponents of gene therapy have tried to block the National Institutes of Health from conducting experimental trials. These critics charge that manipulation of human genes is a "slippery slope" that will inevitably lead to highly questionable practices, such as the creation of humans with extraordinary strength or intelligence or a predominance of males.

Our new ability to manipulate genetic material is raising a number of other ethical and social dilemmas as well: Who will receive gene therapy and who will not? How will the uses of gene therapy be regulated? How will the information obtained from genetic tests be used?

This is not the first time the pace of technology has exceeded society's ability to manage the moral implications of a discovery. This time, however, it will primarily be individual people, not governments, who must make the critical decisions. These decisions should not be made in fear and ignorance.

A recent poll by the March of Dimes Birth Defects Foundation found that 68 percent of Americans know almost nothing about genetic testing, and 87 percent know nothing about gene therapy.

It is crucial that the American public become better educated about this powerful new science. Public debate is needed, and the teaching of genetics must be strengthened at every level of the American educational system, beginning in primary school.

The potential to help Christopher and children like him is at hand. If basic research and clinical trials progress as expected, it will not take more than a decade or two until we find incredible new cures and treatments for people with genetic defects. These advances will present society with complex social and ethical issues. The more informed people become about gene therapy, the better they can speed its potential toward reality.

November 8, 1992

Richard B. Johnston *is senior vice president for programs and medical director of the March of Dimes Birth Defects Foundation and adjunct professor of pediatrics at the Yale University School of Medicine.*

★ ★ ★

Driving to a Safer Future

A. Ray Chamberlain

If driving makes you nervous, especially during winter months when conditions can be treacherous, just consider some trends that could make motor vehicle travel even more dangerous in the future:

- Roads will be more congested, making driving increasingly difficult and motorists more irritable.
- Trucks will be bigger.
- Roads, bridges and the rest of the infrastructure will be older.
- Aging of the population will create new safety problems.
- Cars equipped with futuristic equipment like "talking maps" may distract a driver's attention unless implemented safely.

These and other factors could push the number of deaths from motor vehicle crashes well above the current total of 46,000 annually. In the past, transportation experts have held down fatalities by developing innovations such as safety belts, breakaway highway signs and, more recently, airbags

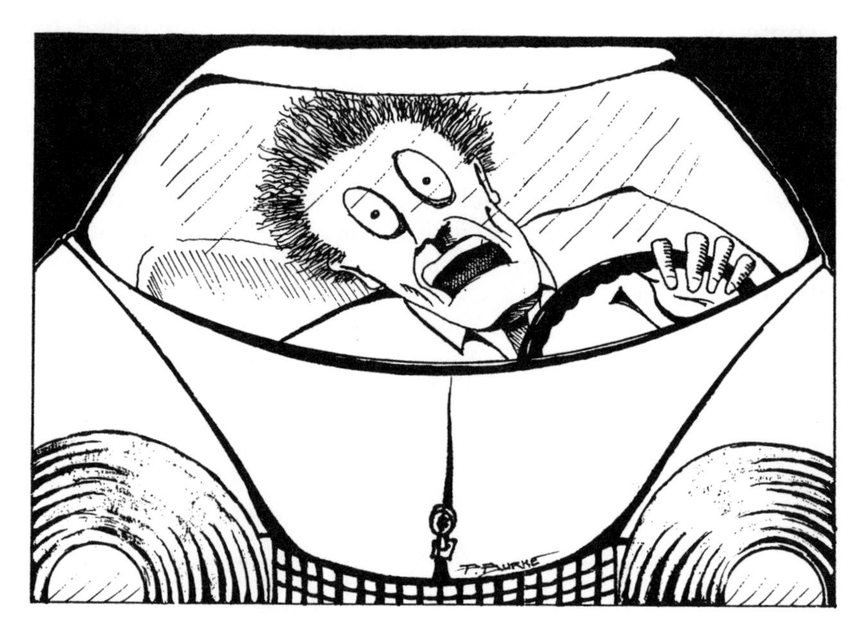

Drawing by Patrick Burke
The Connecticut Post, Bridgeport

and anti-lock brakes. Aggressive efforts to reduce drinking and driving also helped. Although motor vehicle traffic more than doubled since the mid-1960s, the number of deaths per mile traveled plummeted by nearly 60 percent.

But this remarkable progress now appears unsustainable. I led a committee of the National Research Council that recently examined highway safety research, and we found far too little being done to develop the safety innovations of the future. The federal government and the states slashed spending on highway safety research during the 1980s to just $70 million annually, or 30 cents per American. Automobile manufacturers and others also perform safety research, but the overall effort is insufficient.

American motorists deserve better. Fiscal austerity prevents public officials from spending the sums this major public health problem justifies. Yet a great deal could be accomplished by a modest increase in funding focused on a few topics that promise significant results. Our committee identified six in particular that warrant immediate attention:

Crash avoidance. Why do people make mistakes and have crashes? That is a question of growing importance as the population ages and the skills of drivers change. Driver error, sometimes caused by substance abuse, is the major factor contributing to motor vehicle crashes. We need more "human factors" research to learn how to create friendlier vehicles and highways for people with serious — but often predictable — limitations.

Occupant protection. As cars get smaller and trucks bigger, it becomes more important to understand exactly what happens to vehicles and people in a crash. "Biomechanics" research can provide this information.

Safer highways. The curve of a road, the width of a lane and other factors all affect safety. But which highway designs and traffic engineering improvements provide the biggest benefit? We need to learn more. Over the next several decades, major portions of the highway system will be rehabilitated or reconstructed, providing a unique opportunity for making roads safer.

Post-crash acute care and rehabilitation. Emergency medical care is likely to become more difficult as congestion worsens and the number of elderly crash victims increases. New ideas are needed to provide rapid access to crash sites and to improve trauma centers.

Management of highway safety. More older drivers and increased highway congestion also challenge us to develop better licensing programs and law enforcement efforts. Improved methods for screening drivers, for example, might enable states to tailor driver licenses to people's individual capabilities — for example, allowing daytime driving only. Technologies such as automatic speed recording devices could automate some traditional enforcement methods.

Driver information and vehicle control technologies. Automobiles soon may be equipped with systems that warn drivers of unsafe conditions, proximity to other vehicles or even the driver's own impairment. Changeable message signs on the road, meanwhile, will provide real-time information about highway conditions. These information systems could reduce the risk of crashes. But, if poorly designed, they could overload some drivers. It is essential, while these devices are still on the drawing board, that their safety impact be examined.

A research program on these specific topics would require additional federal funding of $30 million to $40 million annually — a tiny amount when compared to the $70 *billion* in medical expenses, lost wages and property damage caused each year by motor vehicle crashes. Anyone unpersuaded by that comparison might consider that 4 million people will be injured in vehicle crashes this year. We owe it to ourselves to reduce this dreadful toll.

January 20, 1991

***A. Ray Chamberlain** is executive director of the Colorado Department of Transportation.*

* * *

New Priorities in the Heavens

John A. Dutton

Once the source of national pride and astounding achievement, the space program has been an invisible issue in this year's presidential campaign. The program was in the news recently because of the change in leadership at the National Aeronautics and Space Administration (NASA). But the candidates have said little about the fabulous opportunities and painful choices awaiting America in space.

The nation needs to re-examine the rationale for its civilian space program. Today's realities are markedly different from those of the past. With the Cold War ended, economic and technological competition among nations is increasing and will determine world leadership. American students are performing poorly in science and mathematics; too few choose careers in these fields that shape our future. With the national budget in deficit, we lack the funds to do all that we could in space, including a host of valuable scientific missions, a space station, a crewed mission to Mars or an aerospace plane to replace the space shuttle. We must choose where in the heavens to set our sights.

We should set them on science, on information and new understanding. Two years ago, a presidential advisory group urged that scientific research be moved front and center in the civilian space program. More recently, a committee that I chaired for the National Research Council agreed that a focus on acquiring information and developing understanding about our Earth and the universe could reinvigorate the space program while supporting other national needs, including the development of powerful capabilities for handling information.

So far, our discoveries in space have revealed unexpected complexity and sharpened our understanding of the universe. We have found new moons and rings around giant planets, seen evidence of black holes, observed the earliest stages of star formation and studied cosmic radiation lingering from the Big Bang. We have searched for life on other worlds and learned much about our own planet, including weather patterns, the detailed mosaics of the surface and the atmospheric ozone hole.

Scientists are planning missions that promise even greater knowledge in the future. Some missions will examine global warming and other environmental problems on Earth. Some will reveal new aspects of the planets, the dynamics of our Sun and the evolution of the universe. We will learn how physical processes react to the microgravity environment of space. To undertake long journeys such as a trip to Mars, we still must determine the effect on humans of living in a weightless environment for several years. The heavens beckon with the promise of undreamed rewards.

To reap them, we must decide what to do. Our committee did not take a position on the controversy between manned and unmanned missions, nor on any specific missions or other NASA programs. Rather, we concluded that focusing on the return of information and creation of enhanced understanding would strengthen all activities in space and enhance the benefits of our national investments. We must set clearer priorities and then follow through with an effective and well-managed program.

NASA cannot afford to complete all of the projects already begun or approved. The scientific community must develop mechanisms to help officials choose among valuable but competing scientific opportunities.

After 30 years of experience in space, certain principles for managing an effective program are clear. Costly missions must be balanced with ongoing support for the scientists and students across the country who quietly do much of the work of space science, converting data into understanding, finding the pathways to be explored tomorrow. Diverse means for reaching space are essential, and both launch vehicles and the role of humans should be matched to mission requirements. We need to find ways to reduce the costs of space research. Once we have started valuable projects, we should finish them.

Our success in space research has brought us many benefits: stimulating young citizens, enhancing national prestige and fostering public pride in national accomplishment. By carefully focusing our efforts, we can increase our gains from space and bequeath still more secrets of the universe to future generations. Although we have an urge to explore new domains, an even more fundamental human imperative is to expand knowledge and know the unknown. Ultimately, that is why we loft our spacecraft into the heavens.

March 29, 1992

John A. Dutton *is dean of the college of earth and mineral science at Penn State University.*

* * *

Reaching for the Answers in the Stars

John N. Bahcall

Imagine looking for a firefly 100 miles away that is glowing next to a brilliant searchlight. That is how hard it is to detect the reflected light from a large planet orbiting a star other than our own Sun.

Astronomers such as myself have other ways of spotting planets. We can study a star to see if it is wobbling slightly, which suggests that nearby planets are exerting a gravitational tug. New telescopes can detect such wobbles and reveal the possible existence of planets 500 light-years away — an astonishing distance.

Discoveries like these may change how we humans think about ourselves, our Earth and our place in the universe. Most Americans know more about the stars in Hollywood than about those in the sky. Practically the only thing they have heard recently about astronomy is the initial troubles with the Hubble Space Telescope. Yet a team of more than 300 scientists that I led for the National Research Council reported this past week that astronomers are on the verge of findings that could transform our collective self-image.

In the coming decade, astronomers will address fundamental questions of nature. How and when did galaxies form? What is the fate of the universe? Are we alone? We humans have pondered questions such as these for millenia.

Anyone who enjoyed the recent "Civil War" television series might consider how much we have learned about the heavens just since Lincoln's time. We now know that stars form out of gas clouds and die either in quiet solitude or in spectacular explosions. Billions of stars are grouped into galaxies that stretch as far as the largest telescope can see. The universe itself apparently was born in a violent explosion some 15 billion years ago.

During the 1980s, we learned that radiation from most of the matter of the universe has gone undetected. We still do not know where — or what — it is. Giant black holes apparently exist in the centers of some galaxies and quasars. Huge neutron stars emit radio pulses that are more regular than the best clocks made by humans.

Still greater advances lie ahead. The country's largest telescope will open soon in Hawaii. New detectors at this and other telescopes will capture light from the heavens with exquisite sensitivity. Combined readings from widely scattered telescopes will yield images much sharper than those from any single site.

Even people uninterested in science have reason to look forward to what we may discover as a result. Astronomy provides benefits that extend far beyond the observatory. Imaging techniques developed for radio astronomy, for example, are now used in medical CAT scanners. X-ray detectors in telescopes led to baggage scanners in airports. Studies of the heavens taught us about environmental problems on Earth, such as ozone depletion and global warming.

Choosing which opportunities to pursue next is not easy. Budgets are so tight that our committee had to set stringent priorities, recommending only about one in ten promising initiatives for funding. The biggest need for "Earth-based" research, we felt, is to strengthen the day-to-day infrastructure. Many observatories have suffered a decade of neglect, and young university researchers are struggling for support. Both need help. Up in space, less costly but more frequent missions are needed to respond to new scientific ideas and technological advances, and to train more young scientists.

We also urged support of four exciting new major telescopes. One of these, a space-based telescope operating at infrared wavelengths, uses camera technology pioneered by the U.S. military for detecting incoming missiles, such as Scuds. It will be more than a thousand times more sensitive than comparable telescopes on earth. A radio telescope in the Southwest and an infrared sensitive telescope in Hawaii will bring distant galaxies into much clearer view. The United States might even begin placing telescopes on the moon if it builds a lunar base.

Other opportunities abound. Up in the night sky, the answers to questions that have mystified mankind for millenia are finally within our grasp. Now that we have the chance, we need only reach a bit farther and grab them.

March 24, 1991

***John N. Bahcall** is president of the American Astronomical Society and professor of natural sciences at the Institute of Advanced Studies in Princeton, N.J.*

★ ★ ★

Abolishing Long-Range Nuclear Missiles

Sidney D. Drell

Although the United States and the former Soviet Union have made remarkable progress on arms control, the scariest weapons of the Cold War remain in thousands of missile silos and submarines. It's time to start thinking seriously about eliminating the long-range ballistic missiles that can hit their intercontinental targets in less than 30 minutes.

Extraordinary progress between the two nuclear superpowers makes this once-Utopian idea plausible. The growing threat of long-range ballistic missiles proliferating among other nations makes it compelling.

By early next century, several new countries may become capable of delivering weapons of mass destruction over long distances. A rogue state could fire — suddenly and with little warning — a long-range nuclear missile at one of our cities. We need a world order free from any nation's having this capability.

Politically, we will enhance our ability to dissuade others from deploying such weapons by agreeing to forego them ourselves. The recent experience with Iraq illustrates the difficulty of detecting covert efforts to attain nuclear capability. Yet long-range ballistic missiles are so large that it is impossible to hide their testing and deployment from current detection technology.

Banning strategic missiles would not require giving up our entire nuclear capability. Rather, it would put a greater burden on our strategic bomber force. This has many advantages. Slow-flying planes require hours rather than minutes to deliver their nuclear devastation. They are far less threatening or useful for first strikes and can be recalled in the event of a misunderstanding or accident.

To maintain its security, the United States would have to ensure that its bomber force could survive an attack. There is a strong basis for believing this could be done. Without long-range missiles, no enemy would have a chance of destroying our dispersed bomber fleet. Equally important, our bombers are equipped with Stealth technology and the ability to fire nuclear-tipped cruise missiles from 1,500 miles away. Experience in the Gulf War has added to our confidence in their effectiveness.

With the disappearance of long-range ballistic missiles, there no longer would be any rationale for a costly, futuristic, Star Wars type of missile defense system. However, cooperation in space exploration would be required to ensure that no country uses acceptable space activities as a cover for military programs that circumvent a missile ban.

Difficult questions remain, most notably: How do we get there from today's force structure? Doing so won't be easy; it probably will take a decade or more. Abolishing strategic missiles would cause tremendous disruptions in our force structure, far greater than those occurring in the current scaling down of the military. It would mean the end of our nuclear triad of submarines, bombers, and land-based missiles, which was developed to meet the former Cold War threat.

Once we have satisfied ourselves on technical grounds that a strategic bomber force could survive and that cruise missiles could penetrate to assigned targets, there would be no strategic reason for continuing to deploy long-range missiles. Politics would be the only remaining major obstacle to their elimination. Not the least of the difficulties would be getting the other powers with long-range nuclear weapons to go along.

In his State of the Union address in January, President Bush proposed cuts that, although large, still would retain a nuclear advantage for the United States. In particular, he called for eliminating all land-based missiles with multiple warheads, the category in which Russia is strongest. Simultaneously, he proposed only a 30 percent cut in the area where the United States is supreme — submarine-launched missiles with accurate, multiple warheads. The problems of balancing these disparate missile forces would end if we got rid of all strategic missiles.

The once-visionary idea of abandoning these weapons — called "fast flyers" by President Reagan when he first proposed such a ban at the Reykjavik summit in October 1986 — may not be achievable in the near term. But as our attention turns from the former Soviet threat to nuclear proliferation elsewhere, it is an idea that demands consideration. It may be possible once again to have a world free of these frightening missiles.

April 19, 1992

***Sidney D. Drell** is professor and deputy director, Stanford Linear Accelerator Center, Stanford University. This article is adapted from a longer version in the Spring 1992 edition of* Issues in Science and Technology.

* * *

Angling for a New Food Source

Robert B. Fridley

Visit the supermarket and you may wince at the price of shrimp, scallops and certain other seafood. The consumer price index for fish has outpaced that for other foods, largely because consumer demand has grown faster than the supply.

But suppose there were a way to increase the supply significantly. Shoppers might get lower prices. Fishing crews, many of which are hurting financially, would get more business. It's a great idea — except that the world's prime fishing grounds already are at or near their maximum sustainable yields. So where would more fish come from?

We should raise them ourselves. That is what the United States does for other foods, from beef to broccoli, better than any other country. Yet although the average American eats 24 percent more fish than a decade ago, we still produce most fish essentially as we have for centuries — by catching what nature provides.

A National Research Council committee, which I chaired, concluded recently that it is possible to supplement natural yields dramatically by raising finfish, shellfish, crustaceans and seaweeds. Some nations do so with great success. Unless we start doing the same, and soon, we risk losing this lucrative market, just as we have fallen behind in the production of textiles, consumer electronics and other goods. Learning to raise fish also would help us enhance natural fisheries.

According to the Food and Agriculture Organization of the United Nations, world fish production in 1988 was 98 million metric tons. Of the 75 million metric tons consumed directly by people, nearly one of every five fish was provided by aquaculture.

This impressive percentage has been growing abroad. But for all of its scientific know-how, the United States contributes just 2 percent of the world's aquaculture production. China is the leader with nearly half of the total. The vast majority of U.S. marine aquaculture production is devoted to a single shellfish — oysters.

Drawing by Bob Rush
The Oshkosh Northwestern, Wis.

Freshwater farming of catfish, crayfish, trout and other species has expanded rapidly in the United States. In some parts of the country these products are now common. Not so with fish farming in saltwater. Barriers to marine aquaculture have been the high value of ocean and coastal space; environmental concerns about animal and feedstock wastes and about the transfer of diseases with wild stocks; and objections by some boaters and fishermen to net or cage installations. Other people say the installations are unsightly.

All of these concerns are solvable, however, and there are large benefits in pushing ahead with marine aquaculture where appropriate. Expanding our efforts in this field could create new jobs, provide a reliable source of seafood, augment threatened fisheries and fish species, and reduce our dependence on imported seafood.

Private companies and entrepreneurs in the United States cannot accomplish this on their own. Our committee concluded that their efforts are being constrained by a regulatory and policy framework that is far too complex and restrictive, and by knowledge and technology that remain too limited.

Suppose you want to begin raising fish offshore. Not only do you need to meet the usual business needs of obtaining financing and the like but you also must comply with federal, state and regional rules that are unclear. Getting a permit can be costly and time consuming. You also must assemble the necessary knowledge, skills and technology. It's no wonder many aspiring entrepreneurs simply give up or go bankrupt.

The permit process must be clarified and streamlined. People who want to undertake marine aquaculture should not face a regulatory maze. Making marine aquaculture a recognized use of the Coastal Zone Management Act, for instance, would stimulate states to include the raising of fish in their coastal management plans.

A modest national research program could provide better methods for raising fish in an environmentally sustainable fashion. Creation of an expanded biological and engineering knowledge base would spur businesses from Maine to Hawaii.

In other words, with a little more scientific and bureaucratic support and a little less red tape, we could ease the pressure on fragile ocean fishing areas while satisfying consumers nationwide. As the demand for fish grows, so must the supply. Mother Nature alone cannot provide all of the fish that consumers want. It's time we woke up and smelled the chowder.

August 16, 1992

Robert B. Fridley *is executive associate dean, College of Agricultural and Environmental Sciences, University of California, Davis.*

10

THE SCIENTIFIC ENTERPRISE

Scientific Openness vs. Litigation Secrecy

Frederick R. Anderson

Is our legal system limiting the ability of scientists to conduct research that alerts Americans to environmental hazards and other dangers? Litigation necessarily requires a measure of confidentiality, yet several recent cases have provoked scientists' concern:

- After the *Exxon Valdez* accident, scientists initially were pleased that millions of dollars would be spent on research to study the oil spill. But some scientists working on the case were not permitted to publish findings, visit damaged beaches, or consult with scientists hired by the opposing parties.
- Medical researchers documented that a link existed between a toxic chemical leak and a family's cancer and neurological disorders. The company responsible for the leak finally settled the case — after the judge agreed to order the settlement amount, the medical research, and court records sealed from public examination. Even local public health authorities were denied the data.
- Another case involved a possible threat to the privacy of research subjects. A scientist was subpoenaed to testify about research he had published. He was asked to bring his laboratory notebooks, including the names of his research subjects.

At a workshop I chaired recently for the National Research Council, several scientists expressed fear that the traditional openness and impartiality of their research is being impaired by trial tactics, confidential settlements, and sealed court records.

Scientists like to view themselves as pursuers of objective truth that is tested by peer review and freely accessible for the public good. They view courts as poor forums in which to pursue scientific truth because of the adversarial nature of trials and the technical complexity of the issues.

For their part, attorneys and judges often resent scientific aloofness. They are sure scientists understand that society requires mechanisms to resolve disputes and strong advocacy for differing interpretations of fact. They also may ask whether scientists are being disingenuous to accept handsome fees yet not feel bound to help win a case.

The problem is that law and science have fundamentally different purposes. Trials require applying legal rules of conduct, sorting out conflicting views of events, and determining fair compensation. Scientists may be needed to establish cause and measure harm, but the main purpose of a trial is not to expand scientific knowledge.

There are strong justifications for many legal practices that are resented by scientists. Coaching scientific experts before a trial, for example, can help shape testimony to meet legal standards of proof and responsibility. Confidentiality may be needed to protect trade secrets and other forms of "information property," or to ensure that intensely personal information is not revealed improperly. Parties to a lawsuit may have to keep information confidential until they have established a strategy.

Judicial confidentiality orders also can have the beneficial, if seemingly converse, effect of promoting *greater* disclosure of facts during trial preparation and pre-trial hearings. Confidentiality also makes it much easier to settle cases without expensive trials. Settlement is the great flywheel that keeps the legal system going; more than 90 percent of cases end up in negotiated settlements that generally are quicker, less costly, and less contentious than full trials.

I know of many scientists who applaud mediation and other dispute-resolution techniques that keep science out of the courts. Yet most litigants will agree to settlements only if some information is kept secret, simply because no court has made a final determination of cause or liability.

Still, adjustments in this time-honored system may be necessary. Trials that cause publicly funded scientific research to be covered up, sealed settlements that prevent environmental and public health officials from gaining access to information, and litigation that blocks scientists from consulting and performing research all point to a need for reform.

Open science and fair trials both are essential in a complex industrial society. If scientists join in a crusade for public access to data gathered for a trial settlement, the legal process could be significantly impaired. Correspondingly, if judges and lawyers stand pat on limiting access to environmental health data, damage may result to scientific inquiry, the environment and public health.

The conflict between these two vital professions hurts the rest of society and must be resolved. Scientists and lawyers each have a role to play, but they also should be able to craft a reasonable compromise on how to work together.

October 20, 1991

Frederick R. Anderson *teaches law at American University and is counsel to the law firm of Cadwalder, Wickersham & Taft.*

* * *

DNA Typing and the Courts

Victor A. McKusick

In courtrooms across the country, lawyers have been battling over the acceptability of "DNA fingerprinting," or "typing." Some defense attorneys say the technology, which compares a suspect's genetic makeup with that in semen, blood or other samples left at a crime scene, is too unreliable to use in determining guilt. Proponents reply that it is the most stunning advance in forensic science since fingerprinting itself.

Although DNA forensic evidence has been introduced in hundreds of trials, courts have varied widely in assessing its reliability. Most courts find the evidence admissible, but others conclude the opposite. Both sides in a case provide technical experts to argue for or against DNA forensic evidence, leaving judges and juries wondering whom to believe. Vast sums are spent on pretrial admissibility hearings. Meanwhile, defendants, victims and others wait for justice.

There is no reason for this uncertainty to continue. I recently chaired a committee of the National Research Council that carried out the most exhaustive study ever of DNA typing. We confirmed the technique's general reliability. When performed properly, it is capable of providing strong evidence for solving crimes.

Our widely anticipated findings were released several weeks ago and should have helped put an end to this dispute. Unfortunately, an initial article in *The New York Times*, which the paper later admitted was in error, caused a great deal of confusion among legal experts and the public alike. That account said our committee wanted a moratorium in using DNA evidence until laboratory standards have been tightened and the technique has been established on a stronger scientific basis.

That was simply wrong. Our report did emphasize the need for a high level of quality control in collecting, analyzing and interpreting data. Some degree of standardization is needed in laboratory procedures, and a mandatory accreditation program should be established. But this does not mean courts should stop admitting this evidence. On the contrary, DNA forensic evidence is a powerful tool for criminal investigation and one that should continue to be used even as standards are strengthened.

Similarly, there is no reason for courts to throw out prior cases in which DNA evidence was admitted unless there is specific information that a laboratory error or other mistake was made in a given case. As a general matter, courts should accept the reliability of the technology and recognize that current laboratory techniques are fundamentally sound.

DNA typing is valuable not only for pointing to the perpetrator of a murder, rape or other crime. It also can clear an innocent suspect. In fact, about one-third of the tests so far have proved someone innocent, sparing the person any further ordeal.

Experts have differed on exactly how to apply and interpret the technology. Much of the scientific controversy has centered on questions involving population genetics. Should a suspect's DNA sample be compared only with those from people of similar ethnic backgrounds or with samples from the general population? The choice affects the statistical chance of a match between a suspect's sample and one from a crime scene being due to random chance.

We endorsed a conservative approach that gives a defendant every benefit of the doubt. When forensic experts compare a suspect's DNA with the sample from a crime scene, they should check at least five locations in the DNA where individual variations are common. In a recent study of several thousand DNA profiles, only a single pair matched over even three locations. No pairs matched when four or five locations were compared.

To protect further against a mistaken match, DNA profiles should be performed on major ethnic groups and the data used to generate probabilities of matches. Doing so would ensure that the stated odds of a random match are weighted in favor of the suspect. To avoid bias, those doing the analysis need never know the suspect's race.

Congress should create an independent expert committee to guide the forensic community in making the most of rapid advances in DNA typing technology. But the courts, the FBI and others need not wait until the technology is further refined. It has advanced far enough for society to use with confidence in helping to identify who is guilty — and who is not.

June 7, 1992

***Victor A. McKusick** is University Professor of Medical Genetics at the Johns Hopkins University School of Medicine.*

* * *

The Legal Barrier to Life-Saving Drugs

Louis Lasagna

If you are pregnant and suffering morning sickness, there is little your physician can prescribe to help you. Similarly, no vaccines exist to protect people against many diseases, ranging from Rocky Mountain spotted fever to AIDS.

Drugs and vaccines like these are badly needed, but one reason they have not been developed is the chilling effect of product liability laws. Flip through the yellow pages and you'll find lawyers seeking clients who may have suffered drug-related damages. To protect themselves, manufacturers may pull drugs and medical devices off the market — even if nothing is wrong with them. The spirit of innovation that spurs new products is threatened.

The example of morning sickness illustrates what this means for consumers. Although usually minor and short-lived, the nausea and vomiting of pregnancy are never pleasant. Millions of women would welcome a drug to alleviate these symptoms safely.

The only prescription drug ever approved in the United States for this purpose was Bendectin, which enjoyed considerable success until assertions appeared in the scientific literature that it could produce congenital defects. A flood of lawsuits ensued from parents claiming Bendectin had harmed their babies. In 1983, the manufacturer voluntarily withdrew the product from the market.

Proving that a medicine causes physical defects is never easy, but it has been especially difficult with Bendectin. A handful of studies have supported the possibility that the drug really can cause defects, but many more have failed to do so. The question remains unresolved. But no matter. Pregnant women have been left to suffer morning sickness without relief, and other companies are reluctant to enter the market.

Thalidomide is a drug that has been even more maligned than Bendectin ever since it was found in 1961 to be associated with "seal limbs" and other congenital abnormalities. No one would dream of approving thalidomide for general use today. But recently, fascinating new uses for the drug have been discovered. Thalidomide has been reported to be effective in treating diseases ranging from leprosy to rheumatoid arthritis.

Many of these conditions are debilitating, painful and recurrent. Treating them with thalidomide might well be worth the possible side effects, just as cancer patients accept the drawbacks of chemotherapy. Yet given thalidomide's litigious history, what company will be brave enough to try remarketing the drug, even for these specific purposes?

Liability laws have become a serious disincentive to research and development in the pharmaceutical industry. There are many reasons why companies decide which drugs to develop, from the likely size of the market to the duration of patent exclusivity. But a growing part of the equation is the threat of litigation. A 1990 study by the Institute of Medicine of the National Academy of Sciences identified liability laws as a major hurdle to providing consumers with more contraceptive options. And the high legal costs associated with liability have been one factor leading companies to boost drug prices.

While the United States relies primarily on tort law to deal with compensation for drug-related damages, other nations have adopted different approaches. Sweden and Japan, for example, have "no-fault" systems of various kinds. The United States enacted a similar approach in 1986, but only for adverse reactions to mandated vaccines, such as those for children.

The current system serves us poorly. Patients who are harmed by drugs require full compensation, but not from a system that is so capricious, slow and inefficient. Some drug awards are too low while others are unpredictably high.

Some states are experimenting with ways of controlling current excesses. Seven states, for instance, have set outer limits on punitive damages, and five have decided that full compliance with Food and Drug Administration product approval regulations may be used by companies as a defense against punitive damages. Some courts have begun seeking independent expert assistance in sorting out the scientific questions that underlie such cases.

Product liability laws clearly are necessary. The public must be protected against negligence and other wrongdoing. But a system that discourages certain kinds of pharmaceutical research, development and marketing ends up harming the very public it is supposed to help. Until this changes, drugs and vaccines that could improve our lives immeasurably will remain undeveloped and unused.

August 11, 1991

***Louis Lasagna** is academic dean of the medical school at Tufts University. This article is adapted from a longer version in* The Liability Maze, *published by the Brookings Institution.*

* * *

Science, Medicine and Animals

Kurt J. Isselbacher

Anyone who says experiments on laboratory animals are unnecessary should explain how physicians learned to treat President Bush's recent illness. The radioactive iodine he took so physicians could scan his thyroid gland was developed through research on rats and larger mammals. The anti-coagulation drug he received to prevent blood clots has been tested on rabbits, swine and other lab animals. Many other aspects of his treatment for Graves' disease also were developed with animal research.

The same is true of current studies of thyroid and heart problems. A research team at the University of North Carolina Medical School is using rats to learn about hypothyroidism. Researchers at Beth Israel Hospital in Boston are doing animal studies of other thyroid conditions. Studies with animals at Columbia University, Case Western Reserve University, Washington University and elsewhere may help physicians learn about atrial fibrillation, the irregular heartbeat that President Bush experienced.

The President's case is notable but not exceptional. At the age of 3, a much less publicized patient named Charlotte Evert had potentially fatal narrowing of the arteries. She underwent a new procedure called balloon angioplasty that widened her arteries and improved her blood flow. The procedure, developed by a physician using dogs and human cadavers, gave Charlotte a normal life.

Greg Maas, a father of two, underwent chemotherapy to overcome a form of cancer that once was invariably fatal. The drugs he took were tested on mice.

There are millions of examples like these. Anyone reading this article has benefitted from animal research that led to vaccines against deadly diseases, treatment for infections, and virtually every other medical advance in this century.

These stories need to be told because advocates of "animal rights" are threatening the efforts of scientists to develop better treatments not only for thyroid and heart prob-

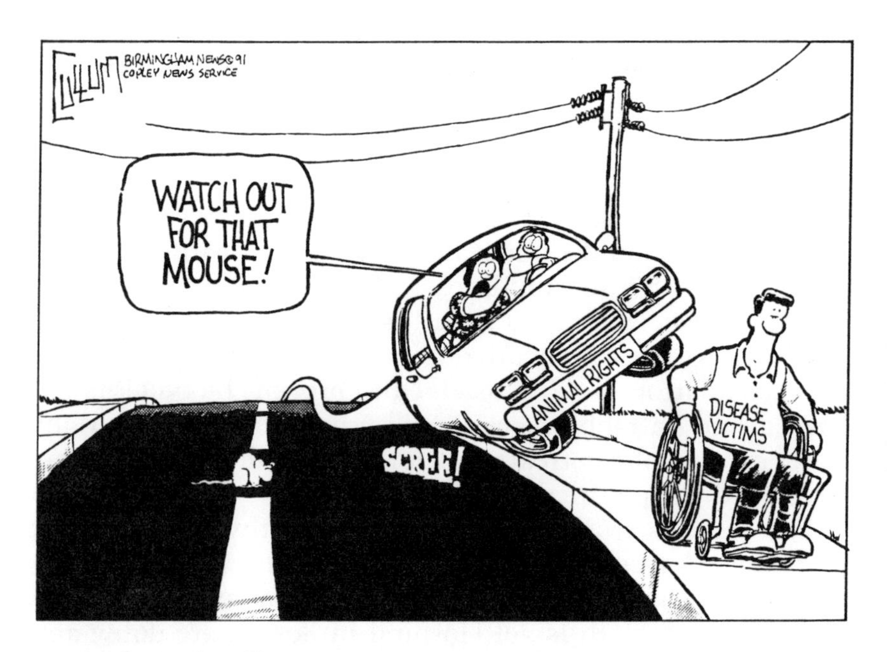

Drawing by Mark Cullum
Copley News Service

lems, but also for cancer, Alzheimer's disease, AIDS and other afflictions. Despite overwhelming evidence to the contrary, these advocates deny that animal research has improved human and animal health.

As a committee of scientists that I chaired for the National Academy of Sciences and Institute of Medicine concluded recently, however, animal research remains an irreplaceable cornerstone of efforts to improve human health. Abandoning it would deny new medicines and cures to future generations.

If humans had chosen a century ago to stop using animals in scientific and medical research, the world would be a very different place today. Many of us are alive — and our parents survived — because diseases were controlled through the knowledge gained from animal research.

This research remains essential, and it is much more limited than its critics portray. The number of vertebrate animals used each year in research, education and testing is a fraction of 1 percent of the number killed for food. Eighty-five percent of the animals are rats and mice. Comparatively few dogs and cats are used, and they come mainly from animal shelters and pounds — which have so many unwanted dogs and cats that they kill approximately 100 for every one provided to scientists.

Researchers would welcome the opportunity to use tissue cultures, microorganisms, computer models and other alternatives in place of animals, which are expensive and inconvenient. Replacements for animals have been developed for some kinds of experiments, and the search for alternatives is continuing. But, for about half of the biomedical investigations carried out in the United States, animals remain essential.

Researchers have an obligation to minimize the pain and distress of lab animals, and to see they are used only for productive goals. On the rare occasions when researchers violate this trust, they should be disciplined. Indeed, "animal welfare" proponents have performed a valuable role in helping ensure that research animals are treated humanely.

But animal rights advocates, who go much farther and argue that scientists should abandon these experiments entirely, owe an explanation to terminally ill children and millions of other Americans waiting for biomedical advances. They should go to the bedsides of these patients and tell them why they are less important than the animals.

May 26, 1991

Kurt J. Isselbacher *is Mallinckrodt Professor of Medicine at Harvard Medical School and director of the cancer center at Massachusetts General Hospital.*

* * *

Preventing Fraud in Science

Howard E. Morgan

It's been tough lately to be a scientist. Incidents involving leading research institutions have given some Americans the impression that science is rife with fraud and other misconduct. That's hardly the case. But as one who led the investigation of a notorious case of scientific fraud several years ago, I am not surprised to see these problems recur.

Most scientists are people of deep integrity. They work long hours searching for cures for diseases, secrets inside atoms, and other mysteries. Although many outsiders regard scientists as a dull lot, researchers generally share a passion for their work. Single-minded in their own pursuit of truth and trusting of colleagues, they have not dealt effectively enough with the possibility of misconduct in their midst.

However, as I learned through my own experience, misconduct is inevitable in any profession. I led a committee formed by the National Institutes of Health to investigate a postdoctoral fellow at Harvard Medical School who fabricated data in nine published papers. The research fellow, John Darsee, was observed fabricating data by a technician. Yet because the institution's investigations lacked rigor and were carried out by persons insufficiently skeptical of Darsee's work, the extent of wrongdoing was not discovered until much later.

The Darsee case received wide publicity. To its credit, Harvard responded by enacting scientific conduct guidelines for its researchers. Other institutions have taken similar steps. But the scientific community needs to do more to fully restore its public credibility.

In doing so, it must maintain the critical distinction between error and fraud. Error is a normal part of science and should not be criticized if it is acknowledged freely. Fraud involves deliberate deception. It takes two forms — the substitution of falsehood for truth, or the selective withholding of truth.

Without clear records and the existence of primary data, this distinction can be blurred. Charges of fraud often hinge on whether published results reflect what really occurred in an experiment. The only satisfactory defense is the presence of a verifiable set of primary data. Sloppy records make for bad science — and for trouble.

When charges do arise, as they have recently, institutions must have effective procedures to deal with them. This need should be apparent to everyone. If an initial review determines that an allegation has merit, an independent committee should be appointed to carry out a prompt and thorough investigation. The members of the committee must be free of any conflict of interest, follow due process and pursue the truth — even if it causes discomfort. Otherwise, both the accuser and the accused remain vulnerable to hostile acts by co-workers, the institution, the news media and government officials.

Fraud is an especially serious problem, but it is not the only issue involving professional conduct among scientists that needs to be addressed. A less publicized problem is determining who gets listed as the author of technical papers. Too often, papers are padded with the names of colleagues who were only marginally involved. Anyone listed as an author should have made a significant intellectual or practical contribution. The submissions letter should define what each author did and certify that all have reviewed the manuscript.

Scientists are eager to be listed on papers because there has been a growing emphasis on quantity over quality of publication when it comes to evaluating people's careers and awarding research funds. One way to remedy this is for institutions to consider only a limited number of publications when reviewing a person's record.

Whatever the specific issue, clear standards of behavior are essential — and institutions must monitor compliance carefully. They need to inform their researchers explicitly about rules involving fraud, use of human and animal subjects, and other matters. And since rules are insufficient, they also should encourage their more experienced researchers to serve as mentors to newcomers, discussing with them the ethical conduct of science. The subtle aspects of how to pursue science are learned not in class but from day-to-day interaction.

Rather than acting defensive about these latest incidents, the scientific community needs to move decisively to reduce the likelihood and severity of future occurrences. After all, if anyone can learn from a recurring phenomenon, it should be scientists.

May 12, 1991

Howard E. Morgan, *director of research at the Geisinger Clinic in Danville, Pa., led the investigation of a publicized fraud case at Harvard Medical School in 1983.*

★ ★ ★

Some of the Toughest Jobs in the World

Norman R. Augustine

Imagine a newspaper ad that proclaims: "Help Wanted. Undergraduate degree required. Graduate degree preferred. Workday 8 a.m. to 8 p.m. Six-day work week. Salary: 30 percent below competitors. Bonus plan: none. Retirement plan: none. Must be willing to move (at own expense). Opportunity to have work critiqued by numerous federal agencies and news media. Supervised by 535-member board of directors. Submit 25-page summary of life and detailed financial statement to White House. Confidentiality not assured."

The description fits many high-level federal government jobs. Finding first-rate people to fill them is difficult enough but particularly so for science and engineering positions.

Some might think of these jobs as plums. In contrast, a

new volume published by the Council for Excellence in Government on the "60 Toughest Science and Technology Jobs in Washington" is titled *The Prune Book*. The Bush administration took an average of nine months to fill these jobs, 50 percent longer than the Reagan administration needed. The next president assuredly will have at least as much difficulty.

You can't run a good horse race without good horses. The same is true for running a government. We need the best scientific and engineering talent available to help repair the environment; provide clean, affordable energy; ensure pure food and drugs; maintain airline safety; research and control diseases; and develop hardware for national security and space exploration. All of these challenges depend mightily upon technological know-how for their resolution.

The government has not suffered a complete erosion of this capability. Yet we cannot assume the problem is not serious. That would be comparable to someone falling off the top of the Empire State Building and observing while passing the 10th floor, "So far, so good."

President Bush has said there is "no greater honor than to labor in government." This is small recompense when considering comparatively low salaries; vague yet harsh conflict-of-interest laws; a burdensome and overly intrusive confirmation process; and post-government employment restrictions that often penalize those who join public service.

Our country properly demands first-rate public employees yet does not pay them first-rate salaries. Sometimes not even second rate. Entry-level graduates earn several thousand dollars less per year than those hired by private industry. Scientists and engineers in mid-level government positions make about half as much as their counterparts in industry.

At higher levels this gap becomes a canyon. For every word he spoke in "Terminator 2," Arnold Schwarzenegger was paid more than $25,000. "Hasta la vista, baby" was more than a year's salary of a senior government scientist working on a cure for cancer. A professional baseball player makes up to $5 million a year — and fails more than two-thirds of the time at bat. An air traffic controller makes less than one percent of that in a year yet can *never* fail.

Financial gain should not be the reason for choosing government service. But neither should financial penalties be the reason talented people must elect *not* to serve. This happens — a lot. During my first tour in government in the 1960s, virtually no one from industry ever turned down a request to serve in a senior government position. Now we have reached the point where many outstanding candidates in the private sector simply are unwilling even to consider public service.

The current conflict-of-interest laws — well-meaning but vaguely written and with severe penalties — have proven to be a lethal combination. And unlike government positions that draw from the legal profession, these scientific and engineering equivalents do little to enhance one's career.

The National Academy of Sciences, National Academy of Engineering and Institute of Medicine recently joined to sponsor a panel chaired by Kenneth Dam, a vice president of IBM. The panel identified common-sense proposals that deserve widespread public support. These include revising conflict-of-interest restrictions; ensuring adequate compensation; and encouraging universities to permit faculty members to take a leave of absence rather than give up tenure.

Fewer than 100 senior positions, among them the "prunes of Washington," oversee the government's role in science and technology. These are the toughest jobs to fill in government. It's in the nation's interest that they be filled by the most talented and seasoned people. A prune truly is a plum — with experience.

April 26, 1992

Norman R. Augustine *is chairman and chief executive officer of Martin Marietta Corp.*

* * *

The Dilemma Behind the Dinosaur Exhibits

Robert M. West

Millions of Americans will visit natural history museums this summer and be dazzled by the exhibits of beautiful gems, mounted animals and ancient fossils. Beyond the dinosaur displays, however, unseen by the public, many of these museums face a dilemma that threatens their vital role in improving our country's appalling level of "science literacy."

Traditionally, most natural history museums considered their main mission to be gathering and studying collections of plants, animals and minerals. Preserving the butterflies came first, public education second.

No longer. With American students being trounced in international science exams, natural history museums are under growing pressure to promote popular interest in science, especially among young people. Many of them are doing exactly that. The Smithsonian Museum of Natural History has developed extensive hands-on discovery areas for children. The Cincinnati Museum of Natural History trains local youngsters to explain exhibits to visitors.

Efforts like these are helping many students to experience the wonders of science for themselves instead of just reading about them in a textbook. The National Research Council and other institutions have called for this kind of informal science education.

With budgets tight, however, many natural history museums now wonder how to maintain their traditional activities while launching public education initiatives. Their research programs are less visible but more important than ever before. With many university departments giving up their collections of "whole organisms," museums are becoming the sole places where scientists can study collections of birds, reptiles and other creatures.

Museum scientists also are carrying out important studies of ecological systems, environmental change, biological extinction and other urgent scientific questions.

So the research responsibilities of museums are expanding right along with the calls for more public education. What is *not* expanding at the same rate is the public funding so critical to many museums. Few private foundations have filled this breach. Changes in tax law have discouraged significant contributions from would-be philanthropists.

The result in many cases has been internal conflict. Museum scientists complain that they are being slighted in favor of watered-down public science. Those who create the exhibits and programs say they cannot provide the public with adequate educational activities. The administration feels besieged to provide more of everything.

Museums cannot sit back and wait for someone to give them more money. They must become more entrepreneurial and businesslike if they are to serve both the scientific community and the public.

While it is difficult to raise funds through science activities, some museums are becoming more problem-oriented in their research and are competing for contracts to do scientific studies. Such efforts can help keep the museums on the cutting edge of research. Yet they also may cause programs to be evaluated according to profitability rather than scientific excellence. Collections and curators might be sacrificed to help balance budgets. This already has occurred in some museums.

Another way museums are coping is by trying to make a profit from their public activities, using television commercials, billboards, contests and other marketing tools. They also have undertaken blockbuster exhibits to pull in huge crowds. A traveling exhibit of bellowing robotic dinosaurs helped several museums to more than double their attendance figures while earning extra money from museum sales shops and cafeterias. Recent exhibits on the Christopher Columbus quincentenerary also have been very popular.

Many natural history museums also are replacing older exhibits with splashy new galleries on topical themes, such as global warming. Museums are developing these exhibits with extensive input from the science, education, exhibits, and marketing and development staffs. They also are expanding their relationships with local schools, providing additional learning experiences for students while gaining extra revenue and good will from government officials.

All of these changes on both the scientific and public fronts are welcome. Natural history museums cannot remain sleepy repositories of insects and minerals. Change won't come easily but museums have no choice. Unless they find a new equilibrium, one that supports their scientific mission while helping ordinary Americans learn more about the natural world, they may become as extinct as the dinosaurs in their exhibit halls.

August 9, 1992

***Robert M. West**, a museum consultant, was curator of geology at the Milwaukee Public Museum. He directed The Carnegie Museum of Natural History, Pittsburgh, and the Cranbrook Institute of Science, Bloomfield Hills, Mich.*

* * *

Too Noisy to Hear the Universe

R. Marcus Price

The electronic din from cellular telephones, garage door openers, remote baby monitors and other devices that use radio waves is making it increasingly difficult for scientists like myself to hear the sounds of the universe. We risk losing the signals from exploding galaxies or even from alien life forms — signals that reach Earth after traveling vast distances.

For decades, our skies have been filled with radio waves used in television and radio broadcasts, aircraft radar, and two-way radios. But in recent years the demand for radio spectrum space has exploded as the miniaturization of equipment has brought us car phones, wireless microphones and many other communications conveniences. Radio signals even are produced, although not used, by computers. All of this technology increases the amount of radio noise in the environment.

That's bad news for radio astronomers, who depend on undisturbed radio waves, much as astronomers who use optical telescopes need light that has not been obscured by urban haze. The radio waves we receive from distant stars and exotic quasars, and even from the very edge of the universe, are extremely weak. Although once very powerful, these signals faded as they expanded into the void of space and traveled millions, or even billions, of years. As they arrive at Earth, they are fainter than a gentle breeze on a tropical isle. It is easy to lose them in the whirlwind of locally generated radio signals.

To hear these faint signals and measure the pulse of the universe, radio astronomers turn their ears to the skies — their radio ears, that is. Giant radio telescopes, some with more than ten thousand times the collecting area of the satellite dishes people have in their backyards, are powerful enough to hear and measure the energy from exploding galaxies near the edge of the universe. They can pinpoint the position of flaring stars more than a hundred times more accurately than an optical photograph, and record the faint radio echo of the enormous Big Bang that heralded the beginning of our universe.

This and other information from radio telescopes translates into more knowledge of our universe and its processes. And the drive to improve radio equipment and techniques helps fuel the explosion of radio technology for consumers. Techniques developed by radio astronomers also are being applied in a wide range of useful technologies, such as devices that detect breast cancer, spot forest fires, guide spacecraft or monitor global environmental changes.

We scientists enjoy cellular telephones, too, and no one suggests that society should stop using new radio communication devices. But the number of radio bands is fixed by nature. It is impossible to manufacture new ones. So we need to protect the bands we have and prevent the rising tide of modern electronic radio noise from threatening scientific research or even telecommunications quality itself. This means using the frequencies we have more carefully. Otherwise, precious information from space will be lost. Astronomers will be forced to spend even more time and effort ensuring that the cosmic data they measure contains no false signals from man-made sources.

I chair a National Research Council committee that has been studying this problem, and we suggest two relatively straightforward ways of easing it. The first is to continue to reserve certain radio bands for scientific purposes. These frequencies should be guarded from commercial applications in the same way that national parks are protected.

The other need is to improve the engineering quality of consumer devices so that their radio waves remain only within designated frequencies. Their radiated power levels should be no higher than required for the devices to work successfully. To keep costs down, many consumer devices now use simple engineering and have inadequate shielding. Their signals spill over into other bands. A major effort should be made to upgrade engineering standards and produce devices of higher quality.

Scientific research and communication technologies can co-exist; the skies are big enough. But as radio electronic devices proliferate, their radio waves must not be allowed to wash away the sounds of the heavens. It is essential that we protect scientific radio frequencies and keep our ears open to hear the answers to some of nature's greatest mysteries.

April 7, 1991

R. Marcus Price *is professor of physics and astronomy at the University of New Mexico.*

* * *

The Blocked Road to Tomorrow's Cures

Katherine Wilson

Anyone who wants cures for AIDS, cancer, diabetes, heart disease and other afflictions depends on biomedical researchers like me. As a new professor at a leading medical school, I am lucky to be among many young scientists who hope to make the great discoveries of the future.

That is becoming agonizingly difficult for us to accomplish. As we try to open our own labs after a decade of advanced training, my friends and I are having unprecedented problems carrying out our research.

The primary source of support for basic biomedical researchers is the National Institutes of Health. Several years ago, NIH began awarding grants for longer time periods. This enabled scientists with proven records to spend more time on research and less time applying for grants. That was a good idea, but the longer grants have soaked up money that used to go to new researchers. The total number of worthy project proposals and the cost of doing research also have increased, further straining the NIH budget.

As a result of these and other factors, the number of grants available to young researchers has fallen substantially since 1988. According to some estimates, as few as one out of 10 new investigators is being funded by NIH. Imagine the effect on your community if most new police officers, or teachers or any other group essential to society suddenly had its primary source of support cut off. That is what is happening to us.

My own situation illustrates the problem. I studied genetics and molecular biology for five years in graduate school, then spent four postdoctoral years doing cell biology research. I landed a job at the Johns Hopkins School of Medicine, and sent my first grant proposal to NIH. It was rated in the top quarter of those submitted and deemed "likely to yield new and important information."

Nevertheless, the proposal was unfunded. When four of

my fellow "new investigators" at Hopkins also were turned down, I began to realize the full extent of the problem. Basic biomedical researchers throughout the country are being affected. A recent conference held by the National Academy of Sciences and the Institute of Medicine attracted Nobel Prize winners and other leading scientists, all of whom expressed deep concern about what this trend means for the future of biomedical research.

My situation was eased recently when the American Cancer Society provided me with limited funding for two to three years, so I do not write out of personal desperation. Nor do I believe that researchers have a birthright to public funds. I have no specific remedy to propose other than giving NIH more money — and I know that the federal deficit is severe. However, private funding sources cannot fill the gap at NIH. Neither can private industry, which appropriately focuses on applied research.

Meanwhile, my colleagues can hang on for only so long. They have the training and the desire to do excellent biomedical research. What a waste of talent if they are forced out! They are full of ideas and ready to pay off society's investment in their training. Many of the young independent scientists are women, a group still underrepresented in science.

The current situation has sent shock waves through the "scientific pipeline," with some undergraduates getting the message that a career in basic research is too risky. We should be sending students a better message about this fascinating and important career.

My friends and I could earn more in private industry than at the university, yet we prefer to unravel the basic mysteries of life and human disease. We would like to continue our work. The basic research we do leads to new knowledge, new vaccines, new drugs and improved medical diagnoses. It provides the foundation for an American export industry — pharmaceuticals — and for our emerging biotechnology industry. Indeed, basic biomedical research is one of the few fields in the world in which the United States still reigns supreme. It is a superb investment of tax dollars.

For many talented members of my generation, however, it also is a personal investment that may now go unfulfilled.

October 7, 1990

Katherine Wilson *is an assistant professor of cell biology at The Johns Hopkins University School of Medicine in Baltimore.*

AFTERWORD

Writing for Newspaper Op-Ed Pages

A Guide To Getting Your Views Published

David Jarmul

You are an expert who has spent years studying an issue of urgent importance to the American people. Now you have a long list of thoughtful recommendations to offer to policy-makers and the public. Or you are a student, parent, business owner or someone else with a point of view that might interest others. How do you get their attention?

One of the best ways is by writing an article for the "op-ed," or commentary, page of a newspaper. An effective article can reach millions of readers, swaying hearts and changing minds. It can reshape a public debate and affect policy. It can bring the author considerable recognition for relatively little effort. But an op-ed article can do these things only if people read it, which means a newspaper must publish it.

This article discusses how to get an op-ed article published. The authors in this book, few of whom had experience writing op-ed articles, succeeded in crafting stories that were accepted for publication by numerous papers across the country. You can do the same. This article focuses on the process of "translating" scientific and academic material for broader audiences, but its advice pertains to anyone seeking to place an op-ed article.

Before racing to your word processor, be aware that the competition for space is intense. Phil Joyce, commentary page editor of *The Philadelphia Inquirer*, says "your chances of getting your unsolicited op-ed piece used are about 30 to 1. With those kinds of odds, you may want to get out of the office and go to the track."[1] At *The New York Times* and *The Washington Post*, the odds are even greater. Several universities that distribute op-ed articles regularly to dozens of newspapers are pleased when even a few papers publish the stories.[2]

Yet some authors do succeed in placing op-ed articles in newspapers around the country, including the most prestigious ones. Why are they successful? How can you increase the odds of getting your own article published?

Understanding the Market

The first step is to look at the situation through the eyes of the person who passes judgment on your article — the paper's op-ed editor. Major metropolitan papers have one or two people working exclusively on the op-ed page. At smaller papers, these duties often are combined with other responsibilities, such as writing the paper's own editorials. In almost all cases, the person reading the article you submitted is an overworked journalist, not an academic.

These editors take their responsibilities very seriously, regarding themselves as the keepers of the microphone at a town meeting. Writes Richard Liefer of *The Chicago Tribune*: "An op-ed page ought to be a place where a wide range of voices can speak to the issues of the day; where controversy can blossom or consensus wilt; where a marginal crackbrain can make a reader sputter over the morning coffee; where four polished paragraphs can bring tears."[3]

"Particularly as more cities have only one newspaper, op-ed pages serve to ensure that a wide range of voices has an outlet that is not filtered through a reporter's word processor," concurs Donna Korando, commentary page editor of *The St. Louis Post-Dispatch*.[4]

Contrary to what some would-be authors believe, editors are hungry for new authors. But they want articles that ring with shock, anger or joy, rather than sounding like scholarly tomes. Without exception, they prefer a gripping personal narrative from a local drug dealer to yet another ponderous analysis of the federal budget deficit.

"The page is at its best when it gives a voice to people who don't usually have one, or who don't usually choose to use it," notes Eric Ringham, commentary editor for *The Minneapolis-St. Paul Star Tribune*. "That principle tends to work against politicians, think tanks and journalists. It tends to favor cab drivers, rape victims and neighborhood activists."[5]

Editors despair of topics that are important but boring. As

Diane Clark of *The San Diego Union* points out, "To some, a debate about disposable diapers is as riveting as one about Germany's reunification; a pro and con on pit bulls as intriguing as mulling over peace dividends."[6] Syndicated writers understand that the best topics are those that readers care about. A content analysis of articles published in 1990 by the seven leading syndicated columnists found more columns on baseball and the death of Greta Garbo than on military spending or developments in Israel.[7]

The Perils of Academic Writing

Dry academic writing rarely works on op-ed pages. Wisdom may suffice in some professional journals, but newspapers also require authors to be timely and, above all, interesting. Readers flipping through the paper on their way to the sports section and TV listings will not stop for anything less. (These readers at least bought the paper instead of turning on the television.) It is not their responsibility to read an article because someone says it is good for them. Rather, it is your responsibility as an author to grab their interest and explain why the topic matters.

If you submit an op-ed piece written in the style of a professional journal, it will be rejected. Kathleen Quinn, who edited articles on *The New York Times* op-ed page for several years, observed that "Most newspaper editors would rather be stranded on a desert island with nothing but a list of the active ingredients in Sinutab to read than so much as glance at another piece of academic prose." Quinn, who says "academic writing stinks," differs from her counterparts at other papers only in her bluntness.[8]

Academics who hear these criticisms sometimes respond by accusing editors of wanting to sensationalize or trivialize complex arguments. As Trudi Spigel, who has overseen a successful op-ed service at Washington University in St. Louis for several years, notes, "There is, to be sure, bashing on both sides. Editors do rail at pompous language, murky arguments, convoluted syntax — as they should. Academics carp at journalists, at editorial pages and editorials. But the fact is that op-ed pages regularly present faculty-written pieces."[9]

Academics and newspaper editors have a common interest in bringing provocative arguments out of the ivory tower to a broader public. The scholars have fascinating ideas; the editors are experienced in "translating" complex material for nonexperts. When they work together, they can accomplish great things.

Far from being vacuous popularizers, most editors are thoughtful people with the authors' best interest at heart. They understand the op-ed medium and recognize that, for better or worse, readability counts more than profundity. As one media critic observed, "Pundit politics are no less corrupt and demeaning than any other kind of politics. It is less crucial to be conversant with the history and culture of any particular field than with the day's headlines in *The New York Times, The Washington Post, The Wall Street Journal,* and the networks."[10]

Practical Guidelines

Here are some basic rules for placing op-ed articles successfully:

- ***Track the news and jump at opportunities.*** Timing is essential. As Kathleen Quinn says about *The New York Times* op-ed page, "When people like Saddam Hussein and George Bush go on the warpath, op-ed editors don't like to hang around waiting to see what next week's mail will bring. And they can't imagine that people will read an article, no matter how wonderful, that bemoans the perennial budget mess when all anybody can think is: 'Does he have the bomb?'"[11]
- ***Limit the article to 750 words.*** That's about three double-spaced, typed pages. Some papers have Sunday editions that run articles as long as 1,200 words. Academic authors often protest that they need more room to explain their arguments. But that's how much space newspapers have to offer and the editors usually are not willing to take the time to cut longer articles down to size.
- ***Put your main point on top.*** You have no more than 10 seconds to hook a reader. One of the most common mistakes newcomers make is using too big a windup before throwing the pitch. Take no more than two or three paragraphs to make your main point, convincing the reader that it's worth his or her valuable time to continue.
- ***Tell readers why they should care.*** Put yourself in the place of the busy person looking at your article. At the end of every few paragraphs, ask out loud: "So what? Who cares?" You need to answer these questions. Will your suggestions help reduce readers' taxes? Protect them from disease? Improve their children's behavior? Explain why. Appeals to self-interest usually are more effective than abstract punditry.

• ***Make a single point — well.*** You cannot expect to solve all of the world's problems in 750 words. Be satisfied with making a single point clearly and persuasively. If you cannot explain your message in a sentence or two, you're trying to cover too much.

• ***Offer specific recommendations.*** An op-ed article is not a news story that simply describes a situation; it is your *opinion* about how to improve matters. Be as specific as possible. Editors will not be satisfied with a call for more research, or with vague suggestions that opposing parties should work out their differences.

• ***Showing is better than discussing.*** One detail or illustration is better than hundreds of words of exposition. Use examples, and then use more examples — ones that readers can understand and care about. Is a government program wasteful? Describe an incident in which $250 of the reader's tax revenues was squandered. That is far more memorable than a concept like "$356 million was lost last year."

• ***Don't be afraid of the personal voice.*** First-person exposition is unusual in academic writing. But with op-ed articles, it can be the best way to help readers understand why you care about the subject. If you are a physician, describe the plight of one of your patients. If you are a physicist listening for signals from alien life forms, tell us the funny questions people ask you. Tracie Sweeney, director of the Brown University op-ed service, says her most successful articles have been from professors who dropped the persona of the dispassionate expert and simply described their own experiences, feelings and views.

• ***Avoid jargon.*** If a technical detail is not essential to your argument, don't use it. When in doubt, leave it out. Simple language does not mean simple thinking; it means you are being considerate of readers who lack your expertise and are sitting half-awake at the breakfast table. Even readers who can identify Africa on a map and who know the difference between home plate and a tectonic plate don't want to wade through difficult prose.

• ***Use short sentences and paragraphs.*** Look at some stories in your local newspaper and count the number of words per sentence. You'll probably find the sentences to be quite short. That's the style you need to use, relying mainly on simple declarative sentences. Search for commas that precede clauses; these often can be made into separate sentences. Paragraphs also should be short. Cut long ones into two or more shorter ones.

• ***Use the active voice***. A sure sign of academic writing is the construction: "It is postulated that . . . ", or "it is recommended that the government should . . . " These are examples of the passive voice, and they leave readers wondering who did the postulating or recommending. Try to use the active voice: ***He*** postulates; ***our panel*** recommended.

• ***No footnotes or citations***. In general, op-ed pages don't use footnotes. There also is no room for authors to thank their dear friends Dr. Erudite and Professor Profundity. If you must mention a colleague, do so quickly near the top, once, without gushing.[12]

• ***Minimize references to your other works***. Your article may be an abbreviated version of a book or report that you just completed. If so, you probably are excited about the longer document. However, most readers won't care. Even after they read your article, they are unlikely to rush out and buy the longer document. So the op-ed article must stand on its own. By all means, refer to the larger work — doing so enhances your credibility — but don't dwell on it. Rather than saying "our committee found this, and our committee recommended that," just make the argument.

• ***Avoid tedious rebuttals***. If you have written your article in response to an earlier piece that made your blood boil, avoid the temptation to prepare a point-by-point rebuttal. It makes you look petty and it's a safe bet that many readers didn't see the earlier article. If they did, they've probably forgotten it. Just mention the earlier article once and then argue your own case.

• ***Make your ending a winner***. Most authors recognize the value of a strong opening paragraph that "hooks" readers. But when writing for the op-ed page, it also is important to summarize your argument in a strong final paragraph. Many casual readers scan the headline, skim the opening column, and then read only the final paragraph and byline. One literary device that often works well at the end is to reprise a phrase or thought made at the beginning, closing the circle.

• ***Relax and have fun***. Many authors approach an op-ed article as an exercise in solemnity. They would increase their chances of publication by lightening up. Newspaper editors despair of weighty articles — called "thumb-suckers" — and yearn for pieces filled with spirit, grace and humor. Readers seek to be entertained and to learn something in the bargain. Obviously, articles on serious subjects must not trivialize their material. But one look at popular syndicated columnists as divergent as Ellen Goodman and James Kilpatrick shows it is possible to combine thoughtful analysis with an engaging style.

Distributing Your Article

The National Academy Op-Ed Service distributes its articles free to more than 300 subscribing newspapers, offering each exclusive rights within its city. In Chicago its articles go to the *Tribune*, in Portland to the *Oregonian*, and so forth. Six prominent newspapers are not on the list because they require op-ed articles on an exclusive basis. These papers are *The New York Times, The Washington Post, The Los Angeles Times, The Wall Street Journal, USA Today* and *The Christian Science Monitor*. Does it make more sense to send an article to one of these papers than to a regional paper? Sometimes yes, but placing an article in them also is much more difficult. And if the article is tied to a breaking event in the news, a rejection means the article appears nowhere. This is because it generally is considered unethical to submit the same article simultaneously to more than one of these papers. Many authors therefore prefer to place articles in local or regional papers which, in many cases, reach large numbers of people and are every bit as important to their customers as *The Washington Post* is to those in Washington. An added advantage of this approach is that many editors prefer local authors to give their pages a hometown feel. Some papers, such as *The Hartford Courant* and *The Fort Worth Star-Telegram*, now supplement their own staff-written and syndicated material almost exclusively with locally generated articles.

Prepare your article typed, double-spaced, with wide margins. List your name, address, phone, and social security number at the top. If you want the article returned if it is rejected, include a stamped, self-addresed return envelope. Learn the name of the op-ed editor and send a brief cover letter, explaining whether you are offering the article on an exclusive or a regional basis. Many papers also welcome a black-and-white photograph of the author, as well as graphics or art. Many op-ed editors prefer being contacted by mail instead of by phone, especially on Thursdays and Fridays when they are completing their Sunday commentary sections. If they do accept your article, be prepared to work with them in the evening. That is when most of the editing is done on morning newspapers, which predominate in the United States.

Which newspaper is most likely to accept your article? Papers in the Midwest are more likely to publish stories on agriculture while those in the Northwest have a special interest in the aerospace industry. But decisions are made by individuals whose tastes are difficult to predict. To complicate matters, an editor who is personally interested in science may feel the op-ed page has carried too many science stories recently. He or she now may be on the lookout for a commentary on teen culture. So if you submit an article on how biotechnology can ease world hunger, your chances for publication are slim. But if it's a story on hungry teenagers, you might have a shot. You'll never know unless you try.

Notes

[1]Quoted in Ciervo, Arthur. "Making It Into the Op-Ed Pages." *Editor & Publisher*, August 18, 1990, 22.

[2]One of the most successful university op-ed services is operated by Anita Goldstein at the University of Southern California. Her service has produced an excellent set of guidelines for would-be authors.

[3]Liefer, Richard. "The Page Is a Vehicle for Intellectual Transaction." *The Masthead*, Summer 1990, 9.

[4]Korando, Donna. "If You Don't Raise Blood Pressure, You're Not Doing Your Job." *The Masthead*, Summer 1990, 10.

[5]Ringham, Eric. "Make a Case or Tell a Story in a New Way." *The Masthead*, Summer 1990, 11-12.

[6]Clark, Diane. "Op-ed Pages Often Take Themselves Far Too Seriously." *The Masthead*, Summer 1990, 6-7.

[7]Croteau, David and Hoynes, William. "Skewed Syndication." *EXTRA!*, June 1992.

[8]Quinn, Kathleen. "Courting the Great Grey Lady." *Lingua Franca*, April/May 1992.

[9]Spigel, Trudi. "Publicize or Perish." *Gannett Center Journal*, Summer 1991, 71-77.

[10]Alterman, Eric. "So You Want to Be a Pundit?" *Utne Reader*, Sept./Oct. 1992, 104-108.

[11]Quinn, "Courting the Great Grey Lady."

[12]This article is much longer than an op-ed article, which explains why it has footnotes.

INDEX

S

T

U

V

W